INSTANT ACTIVITIES
for
MATH
That Kids Really Love!

SCHOLASTIC
PROFESSIONAL **B**OOKS

NEW YORK ◆ TORONTO ◆ LONDON ◆ AUCKLAND ◆ SYDNEY

Edited by Linda Beech. Written by Marty Lee, Marcia Miller, and Joan Novelli.

Cover design by Vincent Ceci and Jaime Lucero

Cover Illustrations by Abby Carter and Rick Brown

Interior design by Ellen Matlach Hassell
for Boultinghouse & Boultinghouse, Inc.

Interior illustrations by Teresa Anderko, Delana Bettoli,
Rick Brown, Drew Hires, and Manuel Rivera.

ISBN 0-590-39947-0

Printed in the U.S.A.

12 11 10 9 8 7 6   3/0

CONTENTS

(continued on the next page)

SECTION 4

Independent Student Activities

SECTION 5

More Skills-Practice Reproducibles

SECTION 6

Teacher's Notebook

Introduction

The importance of tying the teaching of mathematics to the real world is becoming more and more obvious. The National Council of Teachers of Mathematics' standards for what students should know stresses that "mathematics should be connected to other subjects taught in school . . . as well as to situations outside the classroom." *Instant Activities for Math* will help you achieve that goal in quick, fun ways.

Two step-by-step projects show students how and why math figures in putting on a play and planning a pizza party. You'll find lots of tie-ins to other areas of your curriculum. For example, on pages 34–35 there is a wonderful mathematical story from a Middle Eastern culture, on page 53 an art project, and on page 22 a historical link. You'll also find a number of literature links to math such as the problem posed in a famous riddle on page 25 and the story excerpt on page 24 that provides practice in sequencing and organization in an engaging way. There are lots of reproducibles and independent student activities for your math center that connect the real world with estimation, fractions, graphing, problem-solving, and more.

We've also included ideas for setting up a math center, communications to get parents involved, tips about using math journals for assessment, and a bibliography of math books—fiction and non fiction—that make math exciting and accessible to students.

Why not start out with the poster, *Measuring Up,* to get your students involved. Then scan the table of contents to see how all the real-life activities fit into your math curriculum.

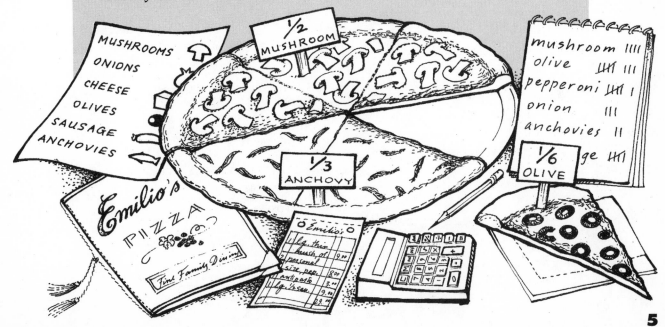

Smart Starts...

...and techniques to use throughout the year

A Egg-Citing Math Center

Egg cartons are excellent math materials—they're free, easy to obtain, and provide endless opportunities for practice activities. Include these egg-carton challenges in your center along with math artifacts such as menus, grocery-store flyers and ads, and the reproducible and independent-student activities in this book.

Dozens of Possibilities

Collect as many cartons as you can to serve as the basic supplies for a math center. Then try these ideas:

Basic Facts Number each cup of an egg carton from 1 to 12 (or 0 to 11). Have students put two beads in the carton, close and shake it. When they open the carton, students determine

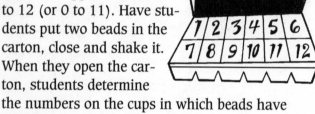

the numbers on the cups in which beads have landed. They then add, subtract, multiply, divide, or list multiples of these numbers.

Fractions Students can show fractions of a set of 12, find equivalent fractions ($\frac{1}{2} = \frac{6}{12}$; $\frac{1}{6} = \frac{2}{12}$...), compare fractions ($\frac{1}{4} < \frac{1}{3}$; $\frac{1}{6} > \frac{1}{12}$...), add and subtract fractions in twelfths, and so on. They can use a numbered egg carton and two beads to form a fraction, then express it in simplest form.

Random Practice Number 12 beads with the digits 0–9, repeating any two digits to make a set of 12. Put the beads in an egg carton, shake it to make each bead land in a cup, open the carton the tall way, and have students use the 2- or 3-digit numbers that form to do column addition or order and compare the numbers.

Probability Students can explore dependent events by putting two or more coins in a carton, shaking it, and then seeing how many heads and tails come up.

Egg Day Challenge students to invent their own egg carton math games. Allow several days for partners or small groups to devise, practice, refine, and write rules for their egg carton games. Then hold a math game eggs-travaganza, in which everyone plays everyone's games.

Activity Cards and Student Reproducibles

Place copies of the independent student activities and student reproducibles in your class math center. (You may want to laminate the activity cards or enclose them in plastic sleeves.) Tell students that each activity card and reproducible features a different math idea. Post a list of the topics where students can easily see them. Assign students to do one or more of the cards in a particular order, or let them select the cards that interest them. Feel free to present any activity to the whole class or to a small group, or use the reproducibles for homework.

On a Roll

Some graphing and sorting activities are tricky to do when there are many objects to manipulate and little room to spread out. But inexpensive vinyl window shades and colored plastic tape can easily become large grids, sorting mats, or other kinds of work surfaces students can use on the floor. Label each shade by describing its math features on a luggage tag so students can tell which shade to take without unrolling them all. You can store the shades on shelves or in a wastebasket or umbrella stand.

Math-Speak

Students feel more comfortable with mathematics when they are familiar with its vocabulary.

Use this activity to help students assess their math vocabulary and identify words that need mastery. Begin by compiling a list of words from your math program or start with the words on this list. Post the words on a large chart pad.

Duplicate the form on this page so that each student has a copy. Review the form and its headings with the class. Then ask students to write the words from the list in the left column. (Suggest that they leave room to add a definition if necessary.) Have students think about their comfort level with each word and then complete the form by making an X in the appropriate boxes. You may wish to model this process with a sample word from the list.

TEACHER TIP
You might also group words on your list, such as words that refer to geometry or words used in fractions.

After students have filled in the forms, invite those who can define each word to do so for the class. Encourage these students to give a sample sentence with the word. Have students who did not check the second box write the definition under the word.

When you have reviewed all the words, have students circle in red those that they are unsure of. Students may want to add these words to a math journal so they can pay more attention to mastering them.

addend
area
average
axis
circumference
common
    denominator
data
diameter
dividend
divisor
equivalent
intersect
median
perimeter
perpendicular
polygon
radius
segment

WORD WISE				
WORD	**I can define it.**	**I can use it.**	**I'm not sure of it.**	**I don't know it.**

Math + Writing = Journals

Math journals fit into today's curriculum as easily and naturally as language arts journals do. Try some of these ideas to blend real-life math and writing.

Math Journal-ism

Math journalists can write about the following questions or others that interest them.

Concept Composition
Have students express their understanding of a math concept. For instance, they might write a concept composition about the meaning of remainders, how fractions and decimals are similar and different, or "What I Know About Percents."

Math Mirror
Encourage students to reflect on their work. Sample questions to stimulate this approach include: What was easy (or hard) about this topic? What did you learn about division? Why did Marcia's answer make sense?

I really like learning about fractions. At the beginning of the year I didn't know how to add and subtract fractions. Now that I understand how to find common denominators, doing fraction problems is much easier. The story about Khalo and the camels helped me understand that fractions can be confusing but they can be very helpful.

What's On My Mind
Ask students to write about how they solved a particular problem or why they chose a particular strategy or solution method. This approach allows students to concentrate on process, not just product.

Research Log
Invite students to jot down questions that might be explored in a survey: What famous person would you most like to meet? How much allowance should a ten-year-old get? What's your ideal travel destination?

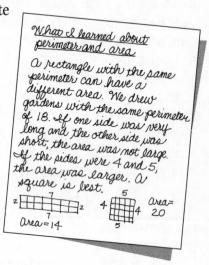

What I learned about perimeter and area

A rectangle with the same perimeter can have a different area. We drew gardens with the same perimeter of 18. If one side was very long and the other side was short, the area was not large. If the sides were 4 and 5, the area was larger. A square is best.

Area = 14   Area = 20

Math-o-Meter
Don't forget the affective side of math writing. Make a "Math-o-Meter" from tagboard, with a movable pointer students can put into position to reflect their attitude toward a particular activity or lesson. After they show their position, they can write about it in greater detail.

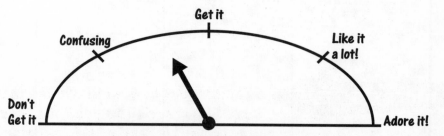

Get it
Confusing
Like it a lot!
Don't Get it
Adore it!

TEACHER TIP
Have students designate a notebook as a math journal. They can divide it into sections for different kinds of math writing, or they can use it as a journal in which to make regular, chronological entries. Individual pocket portfolios can hold noteworthy pieces of work as well as original entries. "Me + Math = ?" on the next page can become the first page in students' math journals.

Name_____

Me + Math = ?

Use this page as the first entry in your math journal.
Complete each statement.

1. The thing I like best about math is _____
_____
_____

2. One thing I don't like about math is _____
_____
_____

3. My favorite math materials are_____
_____
_____

4. When it comes to math, I'm strongest at _____
_____

I'm weakest at _____
_____

5. A math idea I've heard of but don't know much about is _____
_____

I could find out more by _____
_____

More To Try! Think about how math is used in some jobs. Which jobs seem most interesting to you? the least interesting? Write your ideas on the back of this page or in your math journal.

Thinking About Numbers

Get your students exploring some of the purposes numbers serve in their lives. Pose these questions to start a class discussion about numbers: What do numbers tell you? Do you always use numbers the same way? Encourage students to be open in their thinking.

After the discussion, tell students that mathematicians have identified four main uses for numbers:

◆ to **order,** such as 4th grade or Game 5

◆ to **label,** such as 11 Main Street or Route 27

◆ to **count,** such as **15** books or **$879.64**

◆ to **measure,** such as **3:15** p.m. or **60** inches

Point out that some numbers share a purpose. For instance, 5th Avenue **labels** a particular street, but it may also describe the **order** of streets in a city. Post Office Box 17 **labels** a certain mailbox, **orders** boxes in the post office, and **counts** the number of boxes available to receive mail.

Write the current year on the chalkboard. Have small groups use these questions to discuss the purpose of 1998.

◆ Does 1998 count the years that passed since years have been numbered?

◆ Does 1998 measure the passage of time in years?

◆ Does 1998 order the years in this decade, century, or millennium?

◆ Does 1998 label one particular year?

Copy the chart below onto the chalkboard so students can copy it on their own paper. Have students work in groups to list at least eight examples. Encourage students to find examples in all four categories.

NUMBER	WHERE FOUND	PURPOSE

Conclude by presenting the information shown below to the class as a possible poster. Have students identify the purpose of each number used. They can add their descriptions to the chart.

10th Annual Salmon Festival
May 17
noon to 6:00 P.M.
Jake's Fishing Pier
355 Beach Blvd.
Near Exit 32 off Highway 78
Tour 5 fishing boats!
Taste 20 great recipes!
Admission: $3.00
For information call: 555-5453

TEACHER TIP

Duplicate and distribute the reproducible on the next page so students can explore numbers that count, label, order, or measure themselves.

Name_____

Numbers That Name You

There are lots of numbers that label, count, measure, or order information just about you. For example, you have a birthday and an address. Do you have an ID number for a club you belong to? What's your shoe size?

Think about all the numbers that name you. Write these numbers in the person shape. Write some on the wallet too.

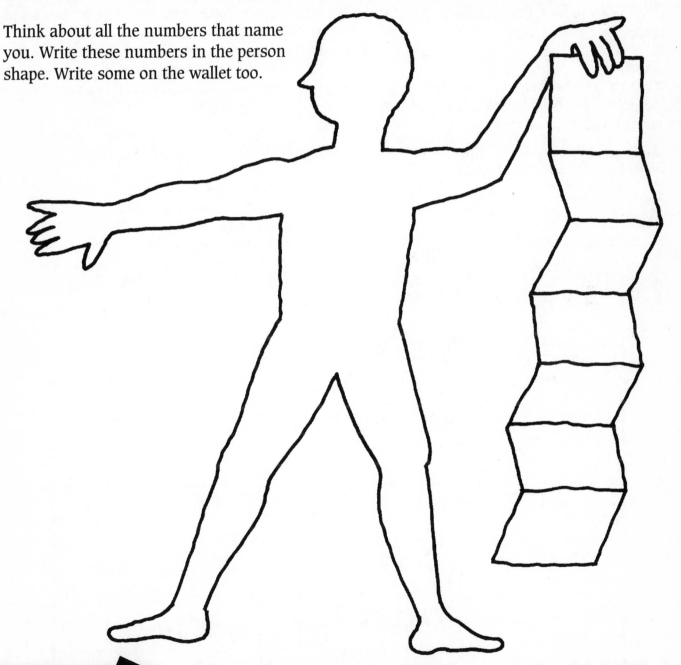

More To Try! → Find out what a social security number is. Do you have one? Why is this an important number?

Math Calendar

A simple outline calendar can provide the basis for a challenging interactive bulletin board that students will look forward to using.

Materials:
large poster paper ◆ colored markers ◆ tacks ◆ pattern on the next page

Steps:
1. On an overhead projector, enlarge the calendar pattern. Your enlargement should take up most of a big sheet of poster paper.

2. Tack the pattern to a bulletin board. Above it add a title: CALENDAR CHALLENGE.

3. Write in the month and year. Determine what day of the week the month starts on.

4. Assign one day on the calendar to each student. Ask students to think of another way to present the numeral for their day. For example, a student who has the second day might write the word duet instead of a 2. A student who has 24 might write a multiplication problem such as 6×4. Encourage students to be as creative as they can. Explain that they can use words, math facts, expressions, geometry problems, fractions, decimals, pictures, or anything else they can think of.

5. Have students add their number substitution to the calendar in the appropriate box.

Calendars as Keepers
You can also use your bulletin board calendar to record daily data. For example, students might keep records of daily classroom waste, attendance, books read, or other information. At the end of the month, have students tally the data and present it on charts or graphs.

CALENDAR CHALLENGE

March

SUNDAY	MONDAY	TUESDAY	WEDNESDAY	THURSDAY	FRIDAY	SATURDAY
		$100 - 99$	$1 + 1$		$8 \times \sqrt{2}$	
★★★★★★	3.5×2	$2 \times 2 \times 2$		$25 + 50 - 70 + 5$		$2 \times 2 \times 3$
XIII		$10 + 9 - 8 + 7 - 6 + 5 - 4 + 3 - 2 + 1$		$4^2 + 1$		$(2 \times 5) + (3 \times 3)$
$50 - 5 - 25$		$(2 \times 2 \times 5) + 2$	$50 - 10 - 9 - 8$		1 dollar − 3 quarters	$723 - 697$
		days in February in Leap Year		1×31		

Duplicate and enlarge this pattern for your bulletin board display.

SUNDAY	MONDAY	TUESDAY	WEDNESDAY	THURSDAY	FRIDAY	SATURDAY

Family Letter

Family members may not think of themselves as mathematicians, but they know they use math all the time. Tap into these daily activities with the following ideas.

◆ Reproduce and send home the letter on this page to alert families to your math plans.

◆ Encourage families to observe all the ways that they use math and add these ideas to your class list (see page 19).

◆ Ask those at home to contribute artifacts to your math resource center. Possibilities include measuring tools, old checks, bills, sales slips, calendars, ads for sales.

◆ Invite parents and caregivers to come to school and have lunch with the class. Seat each visitor at a table for six or eight and have them talk to students about the way that they use math in their daily lives.

◆ Students may not be aware of how many people use math in their work. After sharing the information about a chef on page 26, look for other interesting examples among parents and school personnel. Possibilities include store manager, bank employee, builder, school secretary, architect, dancer, coach. Invite these people to talk to the class and perhaps show some of the math-related materials they use in their work.

◆ Send home the reproducible on the next page and encourage family members to help students fill it out.

Dear Family,

Our class is investigating the real-life aspects of math. You can help by sharing some shopping experiences with your child.

The supermarket or any food store is a wonderful classroom for real-life math. It abounds with price tags and sales signs. You will also find opportunities to discuss discounts, coupons, and weights.

Encourage your child to compare prices of like items, to compute mentally the amount of a discount, to weigh items at the produce counter, or to total and subtract the value of coupons at the checkout counter.

Of course, there are also many opportunities to discuss real-life math at home too. Paying bills, writing checks, figuring out costs, measuring materials to repair, build, make, or bake things all require math skills. We hope you will involve your child in these activities whenever possible.

Many thanks for your help. We think it will add up to some important skills for your child and an appreciation of math as well.

Sincerely,

Name _____

Many a Leg to Stand On

How many legs are in your home?
Become a math detective and find out.
Ask a family member to help you.

1. First, make a guess. _____ **Hint** Not only people have legs. Think about everything in your home—living *and* nonliving—that stands. Also think about legs you can see in photos and pictures. But watch out! Some things have feet but no legs!

2. Now, as you count them all you may want to divide your list into categories, living, nonliving, rooms of the house, and so on. How many legs did you find? _____

How does that number compare with your guess? _____

3. What strategy did you use to find all the legs? _____
_____

4. Where in your house did you find the most legs? _____
Why do you think that is? _____
_____

5. Which room had the fewest legs? _____

6. What was the hardest part of this search? Why? _____
_____
_____

 **More To Try!** Now that you've got a leg up, guess how many feet and how many arms are in your home. Tally both. What do you find?

More Ideas and Tips

Enrich your explorations of real-life math with these suggestions.

What's Numberness?

It's not enough for students to be able to manipulate numbers. They need to have a feeling for numbers—be able to visualize 1, 3, 100, or whatever the number happens to be. This "numberness," comes from real-life experiences with numbers. For example, suppose Natalie is splitting a pizza, cut into six slices, with her brother. Before she has even finished her first piece her brother is reaching for his fourth. She says, "Hey, put it back; that one's mine!" Natalie most likely has numberness for 3, the number of slices in half the pizza. Encourage students to recognize number relationships like this in other events in their lives—whether it's taking attendance or saving money for a special purchase—and to apply these understandings to math problems they encounter.

Math Role Models

Invite parents in to lead a math project and you might scare them away forever. But as Susan Ohanian points out in *Garbage Pizza, Patchwork Quilts, and Math Magic* (Freeman, 1992), if you invite parents in to share something they're comfortable with (that just happens to have math connections), you can help them understand changes in the ways their children are learning math and encourage involvement in reinforcing concepts at home. Quilting, cooking, and architecture are just a few possibilities that work well.

Movable Math Words

Posting math words from areas such as algebra, measurement, and geometry will encourage students to make these words part of their vocabulary—and give them spelling help, too. Here's a giant, portable clipboard you can make for displaying curriculum vocabulary. (See also the suggestions on page 9.)

Materials: large piece of cardboard ◆ giant clip ◆ glue ◆ chart pad

Steps: Glue the back of the clip to the top center of the cardboard. Just clip on the chart pad and you're ready to start adding words. Students who need to consult the list can easily take the clipboard with them—then return it when they're finished.

> **TEACHER TIP**
> Offer to meet with parents beforehand to make some math connections together. This is a good time to introduce terminology students are using that may be unfamiliar to parents.

DEFINING MATH

Math Means...

Carl Sandburg offers one definition of math in this poem. How would students define this important subject?

Arithmetic

Arithmetic is where numbers fly
    like pigeons in and out of your head.
Arithmetic tells you how many you lose or win
    if you know how many you had
    before you lost or won.
Arithmetic is seven eleven all good children
    go to heaven—or five six bundle of sticks.
Arithmetic is numbers you squeeze from your
    head to your hand to your pencil to your paper
    till you get the right answer....
If you have two animal crackers, one good and one bad,
    and you eat one and a striped zebra
    with streaks all over him eats the other,
    how many animal crackers will you have
    if somebody offers you five six seven and you say
    No no no and you say Nay nay nay
    and you say Nix nix nix?
If you ask your mother for one fried egg
    for breakfast and she gives you
    two fried eggs and you eat
    both of them, who is better in arithmetic,
    you or your mother?

—Carl Sandburg

record sports scores
compare data
telephone numbers
shopping
keeping time in music
telling time
measuring

Math Meaning and Me

After sharing the poem with the class, invite students to write their own poems or statements about what math means to them. Post these as a border around your math bulletin board (see page 7) or in your math resource center (see page 36). Students might also copy their statements into their math journals. Stress that what is important is not being the best but being able to use math for different purposes.

So Many Ways

Sandburg touches on many different ways that math plays a part in people's lives. Have students identify these ways. Then work with the class to compile a list of all the ways that people use math. Write the list on adding machine tape and display on a wall. Encourage students to add to the list throughout the year. How many entries can students make?

"Arithmetic" from *The Complete Poems of Carl Sandburg,* Revised and Expanded Edition. Copyright 1950 by Carl Sandburg and renewed 1978 by Margaret Sandburg, Helga Sandburg Crile, and Janet Sandburg. Reprinted by permission of Harcourt Brace & Company.

19

Metric Olympics

Congratulations! Your classroom has been selected as the site for this year's metric olympics. Here's a chance for students to practice their metric skills.

Divide the class into teams. Select a judge from each team to officiate at one of the events. Have each team choose a name, a slogan, and a team color. As a class, decide how events will be scored, if at all, and whether medals will be given. Emphasize that all students will participate. Have a volunteer run around the room once holding a torch (a flashlight?!) and let the games begin!

Here are some metric events to use. Feel free to revise or add to the list or invite students to make up other events. Let students try each event as practice.

Penny Flick

Competitors flick a penny across the floor or table top. The winner is the one whose coin comes to rest closest to 1 meter from the starting line. Judges may need to use millimeters to determine the winners.

Paper Clip Place

From a starting line, competitors try to estimate a distance of 5 meters. They place a paper clip at that spot. The player closest to 5 meters wins.

Jar Fill

Players must try to fill a large empty jar until it holds 1 liter of water. The player who gets closest to 1 liter wins. Provide a gallon jug or large bowl and a graduated liter container.

Getting Warmer

Competitors in groups of four put their index finger into a bowl of water. Everyone guesses the water temperature in degrees Celsius and records the guess on a slip of paper. The judge uses a thermometer to find the water temperature. The person whose guess is closest to the actual temperature wins.

Cotton Ball Golf

Using a ruler as a club, a player calls "fore!" and swings at the cotton ball. The winner is the player who has "driven" the cotton ball the greatest number of centimeters down the classroom fairway. A miss counts as a swing.

Skills-Practice Mini-Lessons...

...to use for specific skill practice

SECTION 2

The Secret of Circles

Introduce this story about area by giving each student a piece of string. Have them tie the ends together to make a loop, then experiment with forming shapes. Ask students to guess which shape would contain the most area. Then read the story aloud to show how knowing the answer helped a princess play a trick.

This ancient Greek legend tells of a clever princess who got an amazing deal on a piece of land big enough to build a city.

Princess Dido asked the king of what is now Tunis to sell her some land. The king refused, but the princess persisted.

"I'm not asking for much," she said. "Just as much land as I can contain with the hide of an ox."

"Well, what can be the harm in that?" said the king. He agreed to the deal.

The clever princess went to work. She and her servants cut an ox hide into very long, very thin strips. Then they sewed all the strips together, end to end, to make one very, very long strip. What do you think Princess Dido did with that strip? She tied the ends together to make a loop.

She took the loop to the king and said, "Here is the hide. Now I would like my land, please." And with that, she used the loop to form a shape that enclosed enough land to build a city. Because she was clever, Dido didn't use just any shape to mark off her land. She chose the shape that enclosed the most area. What shape was that? *(a circle)*

From Strips to Circles

Help students discover the secret to Princess Dido's loop. Start by holding up a piece of paper and asking students to guess how big a circle they could make out of it. Record their estimates, then try this activity to test their answers.

Materials: construction paper or large index cards
◆ scissors

Steps:

1. Have students fold the paper in half widthwise.

2. Show students how to make the first cut—starting at one end of the fold and stopping approximately ¼ inch from the other side.

3. Demonstrate the second cut, starting about ¼ inch from the first cut but at the opposite end, again leaving the last ¼ inch uncut.

4. Students continue cutting this way, alternating sides until they have cut strips all the way across, always leaving ¼ inch uncut. Emphasize that students need to make an odd number of cuts—they may want to plan them out before they make their first cuts.

cut
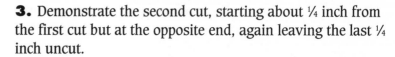

5. For the final cut, have students unfold their papers and cut across the fold, leaving the two end sections (from the first and last cut) uncut.

6. The circle might not hold the city of Carthage but it can enclose quite a few friends!

TEACHER TIP

Have groups discover which string shape encloses the most area. Students might form the shapes with their strings, tape them down, then use uniform counters such as pennies or unpopped popcorn to fill in the enclosed area.

SKILLS-PRACTICE MINI-LESSON / ROMAN NUMERALS

Numerals of Old

Introduce an exploration of the Roman numeral system by asking students to share what they know about ancient Rome. Follow up with this story to take students back to the time this number system originated.

The Web Weaver

Long ago a young girl named Arachne dazzled people near and far with her weavings. So beautiful were her weavings that people said her gift must have come from the gods. Even the goddess Minerva, daughter of Jupiter and a fine weaver herself, admired Arachne's work.

But Arachne refused to thank the gods for her talent. "Not even the gods weave as well as I!" Arachne boasted.

"We'll have a contest." said Minerva. "We'll both weave a picture. You'll see who is best!"

Minerva wove a picture honoring the gods. Arachne pictured gods, too, but made fun of Jupiter. Minerva could see that Arachne's weaving was superior. But she was furious at the way the girl mocked her powerful father.

"You'll weave forever for this!" screamed Minerva, and she flung a potion on the girl. Arachne's arms grew and her body shrank. She became a spider, a form of arachnid. Sadly, Arachne crawled home. The webs she wove were as dazzling as the pictures everyone had admired. Since then, people have appreciated the intricate beauty of the spider's work.

Gift of the Romans

Point out that stories such as "The Web Weaver" are just one example of the legacy left by the ancient Romans. What other examples can students think of? Duplicate and pass out the reproducible on this page to help students discover the use of Roman numerals in their own world.

Where in Our World?

Roman numerals are a gift from the ancient Romans. How many examples can you find?

Find a Roman numeral:

I. **On a clockface**   Place spotted: _____   Time on clockface: _____

II. **In a book**   Book title: _____

The Roman numerals I found are _____

Another way to write that number is _____

III. **On a building**   The numerals that appear on the building are_____

Another way to represent this number is _____

IV. **In a movie or TV show**   The movie or show was _____

I think the Roman numerals tell _____

V. **In a newspaper**   Copy the numerals plus anything that comes just before or after here.

_____

I think these numerals are used to tell _____

Time for a Plan!

Sequencing and organization are skills that challenge students at every grade level—in the classroom and in everyday life. Here are some ways to reinforce these math skills and encourage time management, too.

Share this excerpt from *Anastasia on Her Own* by Lois Lowry. Explain that in this part of the story, Anastasia Krupnik and her family discover the secret for turning chaos into order—and learn a lot about time frames in the process.

Begin by telling the class that Anastasia's mother has a problem: she's disorganized. Anastasia and her father think they can help by planning a housekeeping schedule for Mrs. Krupnik, who also works as an illustrator.

"**N**ow," said Dr. Krupnik, holding his pen poised over a blank sheet of paper, "the thing to do is to divide the day into segments. Each one should be an hour, starting at—"

"Hold it," Anastasia said. "We need a title first."

"Easy. 'Katherine Krupnik's Housekeeping List.' That's the title. Now, if we divide the day—"

"Hold it. I hate that title."

"Why? What's wrong with it?"

"It's sexist. Didn't we once all agree that we should share the housekeeping? We each have a night to do the dishes. Why all of a sudden is it 'Katherine Krupnik's Housekeeping List'? I think that's very anti-feminist," said Anastasia.

"Her father frowned. "You sound like some of my students," he said. "But you're right, I guess. New title:..."

Schedule Solutions

Have students team up to write the schedule they think Anastasia and her father created. Remind students to be sure to give the schedule a new name.

When all teams have worked out a schedule, check the book for the Krupnik Family Nonsexist Housekeeping Schedule. Have students compare their schedules with the one in the book. Invite students to point out possible problems with the book's version. Are there problems with their own schedules? If possible, continue reading to find out what happens, having students read the parts of the story's characters if they like.

Krupnik Family Nonsexist Housekeeping Schedule

7:AM
8:AM
9:A...

Ask students to share organization problems they have. Discuss how a schedule might help solve those problems. Have students plan schedules and test them. Suggest that they begin by estimating how much time they spend on various activities such as getting ready for school or watching TV.

Meaning What?

Solving problems is the goal of real-life math. But real-life problems may have twists and turns that make them hard to solve. This activity helps students sort the information needed to solve a math problem.

"Going to St. Ives" is a famous children's rhyme from England from about 1730. Write it on the chalkboard for a volunteer to read aloud:

Going to St. Ives

As I was going to St. Ives, I met a man with seven wives,
Each wife had seven sacks, each sack had seven cats,
Each cat had seven kits. Kits, cats, sacks, wives—
How many were going to St. Ives?

It's a Riddle

"Going to St. Ives" presents a riddle. Discuss with students what they think the riddle asks. Most will identify a mathematical question: Find the total number of kits, cats, sacks, and wives. Talk about how to find this total. Start a sketch on the chalkboard to guide students to see that the might situation call for multiplication. Have volunteers explain their ideas.

Most riddles have a mathematical slant, even if they lack numerical data, because people approach them by applying problem-solving strategies. Have students list some of the problem-solving strategies they know. *(draw a picture, act it out, use logical reasoning, eliminate extra information, or work backward)*

It's a Trick!

Now tell students that "Going to St. Ives" is actually a problem with extra information that obscures its simplicity. To prove this, ask a volunteer to act out the "I" of the poem. "I" walks on an imaginary road toward St. Ives. Ask other volunteers to pretend to be the man and his seven wives. Have the group walk toward "I." Now reread the final line of the poem: "How many were going to St. Ives?" Regardless of the number of wives, sacks, cats, and kits, it is certain that only *one* person is going *to* St. Ives. The others may be coming from St. Ives, may be crossing the road, or just standing there! Other interpretations are possible. Challenge students to be inquisitive and suggest them!

Work Backward

Pose the famous Riddle of the Sphinx: What walks on four legs in the morning, two legs at noon, and three legs at night? Don't expect students to solve it. Tell them that the answer is a person—a person crawls on all fours in childhood, walks upright as an adult, and then walks with a cane in old age. Once students hear the solution, have them work backward to compare the solution to the riddle to understand its meaning.

A Chef's Story

TEACHER TIP
See page 16 to share a copy of one of John's production sheets with students.

Share this information about a chef who prepares thousands of meals each day—then sends them aboard airlines to feed hungry travelers.

John Troiano went to college to study biology. To help pay for school, he worked as a dishwasher. By junior year he was running the kitchen. John continued his education at The Culinary Institute, a place that trains chefs. Now he runs four kitchens—as executive chef at Borenstein Caterers in Jamaica, New York, a company that prepares airline meals.

People, Plates, and Trays

John starts his day at about 4 A.M. by checking the airline sheets to see how many people are flying that day. John calculates how many trays of food to prepare. "Approximately 24 appetizer plates fit on a tray," explains John. "If 3,000 people are flying one day, I know I'll need 125 trays of appetizers."

John fills out production sheets to order the ingredients for breakfasts, appetizers, lunches, and dinners. When he prepares entrees, he estimates how many people will choose chicken, beef, fish, and vegetarian. "About 75 percent of people flying want chicken," says John. That comes to about 400 pounds of chicken a day!

John puts math to work in almost every other part of his job, too. If he's cooking chicken, he knows it has to reach a temperature of least 165°F before he can take it out of the oven. To prevent harmful bacteria from growing when the chicken comes out, the food goes right into a blast chiller, which quickly cools food to below 40°F.

As food is prepared, it goes onto individual serving trays if it's served cold like salad, or into airline ovens if it's going to be served hot. (These special ovens get plugged in on the planes so that meals can be reheated before serving.) "Depending on the size of the plane, it might take 15 to 18 ovens," says John, "with each oven holding 32 meals."

John ships everything by military time—the time airlines go by. A trucking service delivers the food. Timing is everything. If a plane is grounded because a delivery is late, it holds everyone else up—and can cost the catering company $5,000 a minute! "I haven't been late yet!" says John.

Cafeteria Connections

Obtain a copy of the week's lunch menus from the school cafeteria. Have students form groups and estimate how much of each food they would need for one day. Invite the groups to share how they arrived at their estimates. Then invite the person in charge of ordering food for the school cafeteria to talk with the class about ways he or she uses math in this job and to share the actual amount of each food ordered.

Cookies for 3,000, Please

Invite students to bring in favorite recipes from home—the food they'd like to have served if they were traveling by plane. Next, have them put themselves in John's place and fill out production sheets to order all the ingredients. (See sample, next page.) First, students will need to rewrite the recipes to feed a few thousand. (Decide on the actual number of people flying before students begin.) This is a good time to tie in fractions. For example, what does ¾ cups flour in a standard cookie recipe become when the recipe is multiplied to make 3,000 cookies?

Using a Production Order for Problem Solving

John's day starts with production orders like the one reprinted here. These orders tell John how many people are on each flight, and what meals he needs to prepare, including special requests.

Of course, things change—so the order has extra blanks in case the number of passengers changes or additional special requests are made. Here are a few ways to use the production order with your students.

◆ Display the production order using an overhead projector or divide students into small groups and give each a copy. Ask students to guess what each number tells John. Locate the flight number (018), the type of plane (747), codes that indicate different meals (73—breakfast, 71—lunch, 04—snack), the number of passengers in each class (12, 47, 384), and the number of special meals ordered (right-hand section). Explain that "Kitchen Ready" indicates the time the food must be ready to deliver to the plane (1400 in military time or 2 P.M.). Ask: How is each piece of information useful? For example, people flying first class who order fish might get grilled salmon. But people flying ecomony on the same plane might get filet of sole. Knowing the numbers lets John prepare the right amount of each entree.

◆ What foods would students need if they were in charge of kids' meals for an airline or restaurant? Have groups create production orders with codes for breakfasts, lunches, snacks, and dinners. Have them be sure to note special requests, too. Follow up by having students plan a meal for the class, estimating and recording the amount of each food they'll need. If possible, prepare a simple meal, such as peanut butter and jelly sandwiches, carrot sticks, fruit slices, and pudding. Have students begin by estimating how many loaves of bread they'll need, bags of carrots, and so on. Invite a few guests to share your math-filled meal.

Issued to Kitchen/Bakery/Board Supply/Stores at

KITCHEN READY 1400

	Code	First Class			Code	Business Class			Code	Economy			Code	Crew			Milk	Diabtc.	Fruit Only	Salt Free	Fat Free
		1st	2nd	3rd		1st	2nd	3rd		1st	2nd	3rd		1st	2nd	3rd					
Production Order No. 018	73	12			73	47			73	384			73	40			483	1	1	2	
Flight 018	71	12			71	47			71	384			71	40			Fish	Veg.	Chicken	Dairy	Low Choles.
Date 6/21/95	04	12			04	47			04	384			04	40			1 no butter	14	2 no skin	5 no dairy	
707 747 Freighter																			1 no sauce		
Direct Not																					

Fair and Unfair— Figuring the Odds

When students lose a game, they often complain that the game was unfair. Was it really unfair? Discuss what makes a game fair. Guide students to understand that in a fair game, you may win or you may lose, but all players have the same chance of winning. Point out that you can use math to determine whether a game is fair or unfair.

Make a spinner like the one shown here.

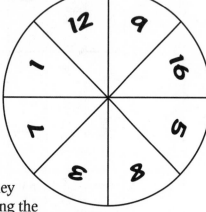

Choose two volunteers to play a game in which they take turns spinning the spinner. Explain the rules of the game:

◆ Player 1 is "odd" and Player 2 is "even."

◆ If an odd number comes up, Player 1 gets a point.

◆ If an even number comes up, Player 2 gets a point.

◆ The first player to get 10 points wins.

Ask students to decide whether this is a fair game. Many will recognize that it is unfair because there are more odd numbers than even numbers on the spinner. Then have the volunteers play a game. Discuss what happens.

Then ask: How could we change the rules to make this a fair game? Students may suggest replacing one odd number with an even number. Others might suggest enlarging the sectors for even numbers so that although there are fewer sectors, they take up more of the spinner. Encourage all reasonable solutions.

Games of Chance

Present the following game situations to students. Have them decide whether each is fair or unfair and explain their decision. If a game is unfair, ask students to change the rules to make it fair.

1. Roll a 1–6 number cube. If a number 4 or greater comes up, you win. If a number less than 4 comes up, your opponent wins.

2. Spin a spinner like the one shown at right. If it lands on white, you win. If it lands on black, you lose.

3. Roll two 1–6 number cubes. If the sum is 8 or greater, your opponent does your homework for a week. If the sum is less than 8, you do your opponent's homework for a week.

28

About How Much

Knowing how to estimate money amounts can be very helpful. Here are some strategies for sharpening students' estimation skills.

Attention, Shoppers

Ask students how estimation is useful to shoppers. They may suggest that an estimate can help determine whether the cashier made an error in entering prices or making change, or whether a shopper has enough money to buy everything in the cart or on a list.

Explain that one useful estimation strategy is known as **front-end estimation.** Front-end estimation is useful for making a quick estimate of prices. Here's how:

◆ Identify the front-end digit—the digit in the greatest place in each number.

◆ Compute using *only* the front-end digits.

◆ Adjust the answer by considering the back-end digits.

Try the following activities to help students understand this strategy.

1. Provide a list of grocery items, all under $10, some under $1. Have students estimate the total.

Example Estimate the cost of this grocery receipt:

Think: Find the front-end total, ignoring amounts less than $1:
$3 + $1 + $0 + $4 + $2 = $10

Adjust by grouping the cents to form dollars. This is about $3.

Since $10 + $3 = $13, the receipt totals about $13.

$3.22
$1.79
$ .95
$4.89
$2.25

$2.35 each

Think: Find the front-end product: 6 × $2 = $12

Adjust to account for the back-end product: 6 × $.35 is about $2.

So 6 × $2.35 is about $14.

2. Provide a problem that involves estimating the cost of several of the same item or of several items having the same price.

Example Estimate the cost of 6 cans of salmon that sell for $2.35 each.

Something to Solve

Have students work with partners to formulate several problems like the two examples. Explain that classmates should be able to solve these problems using front-end estimation.

Number Clusters

Another useful estimation strategy, known as clustering (or averaging), comes in handy for making a quick estimate of numbers that bunch around a common value.

Explain that **clustering** is useful to estimate the weekly attendance at ball games or the total cost of many similarly priced items. Here's how to do it:

◆ Choose an amount around which many numbers bunch, or cluster. Think of this number as an estimated average.

◆ Multiply that estimated average by the number of values in all.

Use these activities to help students understand how to use clustering.

1. Provide the following table of attendance at four football games. Ask students to estimate the total using clustering.

Example Estimate the total attendance for the four games.

2. Provide a problem that involves estimating the cost of several items that are about the same price.

Example Estimate the total cost of the CDs.

GAME	ATTENDANCE
1	58,885
2	56,992
3	60,008
4	61,741

Think: All figures are around 60,000. There are 4 games.

$4 × 60,000 = 240,000$

The total attendance was about 240,000.

TITLE	PRICE
Cat's Chorus	$14.95
The Stapler Singers Stay Together	$13.88
Hamsters' Greatest Hits	$16.19
Deer and Rabbit Hip-Hop	$15.29

Think: All prices cluster around $15.
$4 × 15 = 60$

The total price is about $60.

Cluster with a Classmate

Have students work with partners to make up problems similar to the ones on this page. Have each team present its problems for classmates to solve by clustering.

Key Numbers

Benchmark estimation involves using key numbers to estimate answers quickly. Benchmark numbers are easy to compute mentally. Benchmark estimation is particularly effective for estimating with fractions or percents.

Explain that students should use these two steps to make a benchmark estimation:

◆ Choose a sensible benchmark.

◆ Use mental math to compute.

Try the following activities to help students understand how to use this strategy.

1. Provide a situation that involves combining cooking ingredients. Ask students to estimate a total weight.

Example Estimate the total weight of trail mix made from 1⅚ lb nuts and ⁴⁄₁₀ lb raisins.

Think: 1⅚ is about 2. ⁴⁄₁₀ is about ½.
2 + ½ = 2½

So 1⅚ lb + ⁴⁄₁₀ lb is about 2½ lb.

Think: 54% is about 50%. 279 is about 300. 50% of 300 is 150.

So about 150 Linton students speak Spanish at home.

2. Provide a situation that involves estimating the percent of a number.

Example Fifty-four percent of the 279 students at Linton School speak Spanish at home. Estimate how many students speak Spanish at home.

Have students make up problems like these for classmates to solve using benchmark estimation.

From One Thing to Another

Another way to use benchmark estimation is to begin with a known quantity and estimate with it. Discuss with students how to use benchmarks to estimate real-life situations involving distance, weight, or time. For example, students who know that a basketball rim is 10 feet high can use a mental picture of it to estimate other distances, such as the length of a driveway or the height of the gym ceiling. Similarly, someone who knows the feel of a 15-pound cat can use that sensory information to estimate other weights.

Test-Taking Techniques

Students are likely to take tests in which computations are required. Flexibility with estimation is an excellent approach for students to use when calculating or checking the reasonableness of answers.

In checking answers on a test, students can use whatever kind of estimation best suits a given problem or situation. Here are some helpful tips they should know.

1. Use *flexible* rounding—substitute numbers that are easy to work with. For example, you can round numbers to a given place, to the greatest place, to the nearest dollar, to the nearest whole number, or to a multiple of 10.

2. Use any of the estimation strategies you know—front-end estimation, clustering, or benchmark estimation.

3. Compute with the rounded or adjusted numbers.

4. Use your number sense to adjust results to compensate for underestimates or overestimates.

Guide students through these examples of flexible estimation.

Example Find the product: 44 × 94

Think: Round 44 *down* to 40. Round 94 *up* to 100. 40 × 100 = 4,000 Normally, 94 is rounded to 90, but it makes mental computation easier to round it to 100. Since one factor is rounded up and the other rounded down, the estimate of 4,000 is reasonable.

Example Find the sum: $5.59 + $3.95 + $7.89

Think: Round each amount to the nearest dollar. Add: $6 + $4 + $8 = $18 Since all amounts were rounded up, $18 is an over-estimate. Compensate by subtracting $1.
So the sum is about $17.

Example Find the quotient: 3,827 ÷ 93

Think: Round 3,827 to its greatest place, 4,000.
Round 93 to 100, a convenient multiple of 10. 4,000 ÷ 100 = 40
So the quotient is about 40.

TEACHER TIP

For each example of flexible estimation, have students compute the exact answer. Then compare the estimates with the answers so students can see how reliable the estimates are.

Remind students that estimation will help them determine if an answer is reasonable, not the exact answer itself. Encourage students to be flexible in their approach to estimation to check that their answers make sense the next time they take a math test.

Figuring in Fractions

After reading and responding to the folktale, The Clever Counter, on the next two pages, challenge students' problem-solving skills with these activities.

Which Is More?

In the story "The Clever Counter," Khalo finds a satisfactory way to divide an inheritance of 35 camels among 3 siblings so that, according to their father's wishes, one gets ½, one gets ⅓, and one gets ⅑. The solution, of course, requires an understanding of fractions. Help students understand fractions by first asking them to identify and compare the fractions in the story. Have students explain their reasoning in writing, drawing pictures, too, if they want.

How Did He Do That?

Encourage understanding of the concept of common denominators by taking a closer look at the brothers' problem—and Khalo's solution. First ask: What was the problem with dividing 35 camels among the brothers as the father wished? *(The division results in remainders—½ a camel, and so on.)* Next have students play the part of the story's clever character and in writing explain how he arrived at his solution and why it was profitable for all. As students share their reasoning, encourage them to recognize that 36 is the common denominator closest to the original number of camels— 35—and can be divided into ½, ⅓, and ⅑ without leaving the brothers with parts of camels. Help students to understand, too, why there were 2 whole camels left over ($\frac{1}{2} + \frac{1}{3} + \frac{1}{9} = \frac{18}{36} + \frac{12}{36} + \frac{4}{36} = \frac{34}{36}$).

Add a New Ending

Review Khalo's method for dividing the camels among the three brothers. Ask students to think about other ways to solve this problem. Have students rewrite the second half of the story as though they were Khalo, offering solutions they think are fair and mathematically accurate. Keep in mind that not all answers will reflect traditional division.

Everyday Fractions

Students may not be able to work with camels to learn more about fractions, but they can find lots of examples in the world around them. Post a chart for recording experiences with fractions. Start by brainstorming examples, such as:

◆ It's half past three.

◆ I spelled nine out of ten words correctly.

◆ You can have half of my sandwich.

◆ I'd like two-thirds of a pound of cheese.

Encourage students to recognize and record other examples they hear or use themselves and add them to the chart.

Name_____

LITERATURE LINK

The Clever Counter

In this Arab folktale, you'll meet three brothers who are struggling with a problem and cannot find a solution to it. You'll also meet the man who puts math to work to find a quick and clever solution.

As you read the story, write your comments, questions, predictions, and math ideas in the margin. A few have been done for you as samples.

Dividing camels is different from dividing fruit!

My friend Khalo must be a mathematical genius. I know because one day he found a way to divide 35 camels among three brothers—with the first brother getting one half of the herd! Impossible, you say? Half of 35 is 17½, and who would walk away with half a camel?

Well, here's what happened. Khalo and I were swaying along on the back of my camel. Khalo did not have his own camel so we doubled up, uncomfortable as it was. We were on a desert road approaching Baghdad when we overheard three men arguing.

"No way!"

"You're a cheat!"

"That's not fair!"

Khalo, always interested in other people's business, stopped and asked what was wrong.

Did the father know what problems this would cause?

"We're brothers," the oldest man explained, "and when our father died, he left us 35 camels. Half of them are supposed to go to me, my brother Hamed is to get one-third, and Harim, the youngest, is supposed to get one-ninth."

Then Hamed broke in. "Half of 35 is 17½," he said. "None of the other numbers comes out even, but don't expect me to give up any of MY share!"

"Me either," said Harim.

"Well," said Khalo thoughtfully, "I think I can help you make a fair split. And I'll even throw in my friend's camel so you can start with 36 instead of 35. How's that?"

"Wait!" I cried, as I led Khalo a few steps away. "Are you crazy? Do you want to walk the rest of the way to Baghdad in this desert?"

"Trust me," he replied quietly. "You'll get the camel back. I know what I'm doing. It'll all work out. You'll see."

So I handed him the reins of the camel. If I had to walk, he'd have to walk. And I really didn't think he'd let that happen.

"We now have 36 camels to divide, not 35," Khalo said. "I'm going to make sure that each of you gets the share your father wished."

He turned to the oldest brother. "You would have gotten half of 35—that is 17½. Instead, you'll get half of 36. That's 18, so you come out ahead. Okay?"

The oldest brother nodded agreeably. I could only wonder how I was going to get my camel back if Khalo was being so generous.

He turned to the middle brother. "Hamed, your share was one-third of 35, so you would have had 11⅔ camels coming to you. But now you get an even 12. So you're ahead. Any problem with that?"

Hamed shook his head "no."

Then my friend turned to the youngest, Harim, who was looking just as worried as I felt.

What if Khalo's math was wrong?

"Your father's will leaves you one-ninth of the camels. Out of a herd of 35, that would be more than three camels but fewer than four. But now, you'll get one-ninth of 36—that's four camels even, more than you would have gotten."

Harim smiled.

I didn't. If they're all getting more than they would have, I thought, how can I possibly be getting my camel back?

Then Khalo casually summed up the solution. "The oldest gets 18 camels, the next gets 12, and the youngest gets 4. Add them all up—18 plus 12 plus 4—and you get 34. You all wind up with more than your share, and out of 36 camels there are 2 left over."

I did some quick figuring and so did the brothers. It all added up but seemed impossible!

Before we could say anything Khalo went on: "One of those two camels is my friend's. As you can see, the thought of walking all the way to Baghdad has him a little worried. I'm sure he'll be happy to have the camel returned to him. And it's only fair, I think, for me to have the other because I helped you settle the argument and carry out your father's wishes."

How could anyone argue with that? Khalo picked out one of the best camels. It knelt in the sand so that he could climb up, and we continued in comfort to Baghdad.

Two Step-by-Step Projects...

...to get your students to make real-life decisions based on math

Project 1—Putting on a Play

Project 2—Planning a Pizza Party

Making Math Perform

For this six-assignment project students put on a play or other performance and explore the math behind the scenes.

Getting Ready

Distribute some play programs or performance posters. Ask students to look through them for ideas on the kinds of pre-performance activities that take place, such as raising money, making costumes, props, and scenery, planning lighting, and preparing music. Talk about what other information the programs or posters provide.

Talking About the Task

Conduct a class survey to determine the kind of performance students would like to undertake. Brainstorm different kinds of performances and list ideas on the chalkboard. Students might suggest plays, puppet shows, dances, skits, poetry or story readings. Have the class use the list to vote for the kind of performance they'd most like to put on. Tally the responses and discuss the results.

Have students make a bar graph to show the results of the class vote. Display and discuss the graph and talk about how it shows students' preferences, total number of votes, and other information.

CLASS SURVEY:
KINDS OF PERFORMANCES

play |||| |
poetry ||
story reading ||||
puppet show |||| ||
dance |||

TEACHER TIP

When the project is completed, have students compare the actual time spent with the estimates on their time lines.

FIRST ASSIGNMENT
Behind the Scenes

Tell students that it can take more time than they think to put on a show. For example, a one-hour play might require weeks of preparation. Brainstorm with the class to list all the tasks that would be needed to put on their favorite kind of performance, from inception to curtain call. Form a small group of students for each task. Ask each group to estimate how long, in days, their task will take.

Come together to propose a schedule for putting on the performance. First, have each group report on its estimate. Discuss which times it makes sense to *under*estimate and which should be *over*estimated. Then use the estimates to complete a start-to-finish time line.

Here is the beginning of a sample time line:

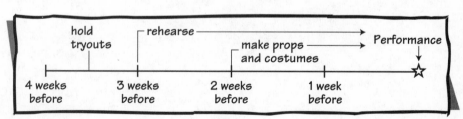

SECOND ASSIGNMENT
Patterns at Play

You can enhance any performances with melody, movement, or rhythm. Invite students to sing or act out signature songs or dances from TV shows or films they know. Then divide the class into small groups. Have each group choreograph a movement pattern or dance, or compose a rhythmic pattern or melodic line.

Show students how to use a system of icons, glyphs (see page 44), colors, or letters to record their creations. For instance, this movement pattern is shown with letters: J/T = jump and twirl; C/S = snap and shimmy; SS = sidestep.

J/T C/S SS SS SS J/T C/S SS SS SS...

This rhythmic pattern uses symbols for stomps (■), snaps (★), and claps (▲):

■ ★ ★ ■ ★ ★ ▲ ■ ★ ★ ■ ★ ★ ▲...

Tunes can be written in standard musical notation, or with arrows or steps that indicate the shape of a melodic line. Display two ways to show the familiar tune "Row, Row, Row Your Boat":

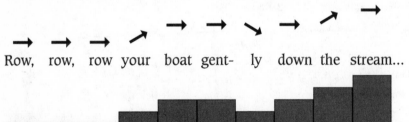

Row,   row,   row your   boat gent-   ly   down   the   stream...

Have groups develop, record, and share their creations. Add the class favorites to your show, as appropriate.

Mirror Movement

Student pairs can apply the geometric concept of symmetry to explore reflecting movement. Divide the class into partners of approximately the same height. Have the partners stand facing each other, separated by about a foot of space. One student is the Mover. The Mover makes simple motions that the partner, as the Mirror, reflects. Movements can be smooth or jerky, fast or slow, large or small. Have partners switch roles so they can compare how it feels in each position.

Encourage pairs to share their mirror movements with classmates. Partners who are adept at mirror movement may want to incorporate it into the performance.

THIRD ASSIGNMENT
Number Crunching

Putting on a play costs money, but it can raise money as well. Have students imagine that their play will be a fund-raiser. Ask them to consider what audience members should pay to attend a performance. List every price students propose. Make an entry for every student. Then work with students to organize the data into a line plot to show the data. Here's how:

1. Draw a horizontal line segment.

2. Mark a scale of numbers that fits your data. (Use the range of your data as a guide.) Include every number mentioned.

3. Make an X for every entry. For instance, if three students named a price of $2, make 3 X's above $2 on the scale.

4. Give the line plot a title.

In this sample, ticket prices range from $1 to $10. Most students said prices of $5.

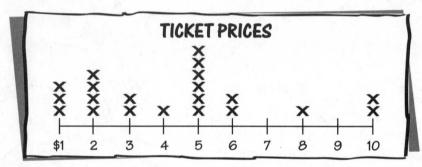

Talk about what *your* class line plot shows by asking questions like these:

◆ Which price did most students choose? How can you tell?

◆ What is the range of ticket prices?

◆ Do prices bunch up around certain numbers? Which ones?

◆ What price(s) shown on the line plot was not suggested?

Use the information on the line plot to set a ticket price for the performance. Then have students estimate projected revenues based on that price.

Raising More

Have students suppose that the projected revenue won't be enough. Discuss ways to raise more money from the performance. Suggestions might include raising ticket prices; giving more performances; using a larger performance space; offering extras, such as front-row seats or attendance at the cast party for an additional fee; or selling food, programs, and buttons. Evaluate each option to find those that make the most sense.

Projected Revenues

ticket price $5

projected ticket sales   50 tickets

projected revenues $250

TEACHER TIP

If students are doing a fund-raiser, repeat the steps you followed under Talking About the Task (page 38) to decide on a cause.

FOURTH ASSIGNMENT
Design Time

Most performances need scenery, props, and costumes. Talk with students about how set and costume designers and prop builders first make detailed sketches, then use their sketches to make the actual items. Discuss what could go wrong if the sketches are unclear or labeled inaccurately.

Divide the class into pairs. Ask each pair to sketch a design for a costume, a piece of scenery, or a prop for the performance. Emphasize that sketches must be drawn carefully and completely on grid paper. Sketches must include all necessary dimensions given in exact measurements, as well as a sequence of steps that the people who will sew or build can follow. In addition, students should describe all the shapes and features in the sketch using precise math language. It may help students to imagine that they cannot talk with the people who will sew or build.

Here is a sample scenery sketch of a cardboard house.

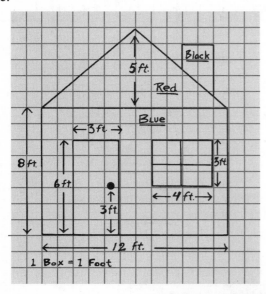

Invite pairs to share and describe their drawings. Post the completed work for students to examine, or include the sketches in a class portfolio.

FIFTH ASSIGNMENT
Getting Seated

Have students imagine that they will perform their play in one of the school's large open areas, such as a gym or cafeteria. To help students develop a seating plan for the performance, duplicate and pass out the worksheet on the next page.

TEACHER TIP
Make a class list of other jobs that require the use of measurement.

Name_____

Put Them in Their Seats

Use the grid paper to make an audience seating plan for your performance. Follow the rules inside the "stage."

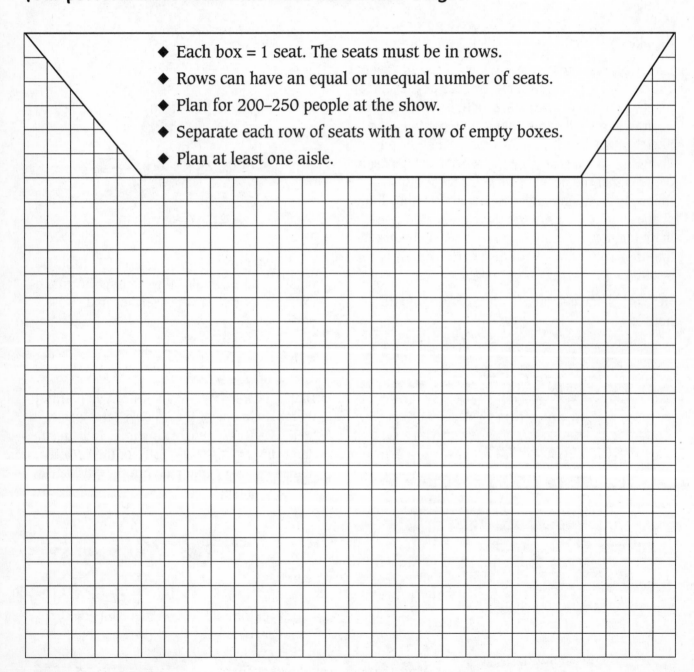

- ◆ Each box = 1 seat. The seats must be in rows.
- ◆ Rows can have an equal or unequal number of seats.
- ◆ Plan for 200–250 people at the show.
- ◆ Separate each row of seats with a row of empty boxes.
- ◆ Plan at least one aisle.

On another sheet of grid paper, design a seating arrangement for a performance with seats around a circular stage.

Sixth Assignment
It's a Wrap!

Now that students have chosen a performance, figured out what to charge for tickets, designed sets, props, and costumes, and planned a seating arrangement, let the play begin!

After the performance, have the groups look back at the estimates they made for the time each behind-the-scenes task would take. Ask them to prepare answers to these questions to compare their predictions with the actual results.

◆ How long did it really take to do the task?

◆ Did you *over*estimate or *under*estimate?

◆ What surprised you the most?

Have each group present its findings. Ask students to explain how they could use what they learned to adjust their estimates for a future performance. Then have them adjust the time lines they made, or make new ones, to reflect their understanding of the actual times things took.

TEACHER TIP

Students can apply their estimation skills to the actual performance. How long do they think it will take? How long does the dress rehearsal take?

What Did We Learn?

Distribute the play programs again. Point out the part that acknowledges important contributions by individual people or organizations. Have students browse through the playbill, now as seasoned producers, to write acknowledgments of all the ways that using math enhances the process of putting on a show. Students can add these acknowledgments to their math journals or portfolios.

Pizza Party

For this five-assignment project students use their math skills to plan and hold a class pizza party.

TEACHER TIP
Be sure to share the information about a chef and his work on pages 26–27.

Getting Ready

Talk with students about the idea of using symbols to give information. Ask them for examples of symbols and their meanings. Discuss the various symbols used in mathematics. Guide students to visualize weather maps, road signs, or pictographs and discuss why using symbols can be effective. Then introduce the term **glyph**—a symbol that stands for information.

Draw an outline of a face on the chalkboard. Include all the features except the mouth. Tell students to imagine a symbol or glyph they can use for the mouth. Explain that the mouth symbol will stand for their favorite food. Emphasize that although glyphs need not look exactly like the item they represent, they should suggest it clearly. In this case the glyph should suggest a mouth and a favorite food. List students' suggestions on the board and invite volunteers to sketch a glyph for each one. Then draw one of the glyphs for the mouth on the face.

Talking About the Task

Explain to students that they are going to have a pizza party, but that it will take planning and preparation. Begin by brainstorming a list of pizza toppings. Be creative as well as realistic—onions, pepperoni, or squid! Agree on glyphs you can use to represent each topping. Create a glyph key by having volunteers draw a glyph for each topping on the list. Then make glyphs for favorite beverages.

cheese · olives · onions · peppers · mushrooms · pepperoni · anchovies · squid

FIRST ASSIGNMENT
Glyphs

Talk with students about how they could use glyphs to show their ideal pizza and favorite beverage. Then duplicate and pass out the worksheet on the next page.

FRUITS

TEACHER TIP
Draw students' attention to the many glyphs used in computer programs.

Name_____

Top This!

Fill the head with glyphs to show what toppings you would put on your ideal pizza. Glyph the glass to show what you like to drink with your pizza. Make a key to show what your glyphs represent.

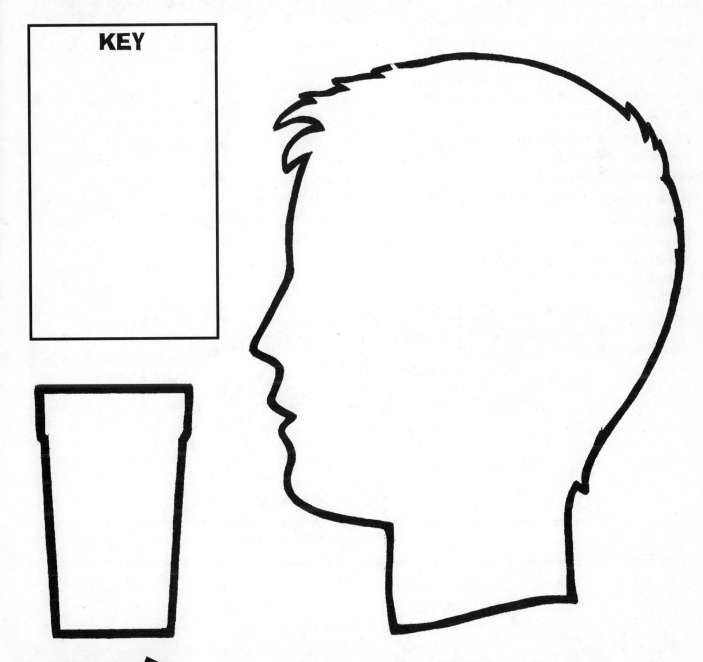

KEY

More To Try! → Tack your finished pizza and drink glyph on a bulletin board. Write a riddle about it, such as: My ideal pizza has only green vegetables and one kind of meat, and comes with lemonade. Which order is mine? Try to solve each other's riddles.

SECOND ASSIGNMENT
Weighted Voting

The Two Toppings

Discuss with students why ordering party pizzas with dozens of toppings wouldn't make sense and would drive the pizza place crazy! Point out that a better idea would be to select students' *two* favorite toppings. List on the chalkboard all the toppings students named for their ideal pizzas.

Vote on a favorite topping from the list. Allow each student only *one* vote. Tally the votes, then ask questions like these:

◆ Which pizza topping is the class favorite?

◆ Which is the second favorite? Just because a topping gets the second most votes, is it necessarily the second most popular topping? Talk about it.

Introduce the strategy of weighted voting, in which each student gives 2 votes to his or her favorite topping and 1 vote for the second favorite. In this system each student actually casts **3** votes (2 + 1). Vote on the pizza toppings again, using the weighted system. Tally the results, then ask:

◆ How do the weighted results compare with the original results? Explain.

◆ Which way of voting gives a more accurate picture of class favorites? Why?

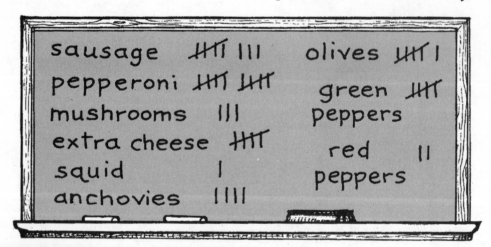

sausage ~~||||~~ ||| olives ~~||||~~ |

pepperoni ~~||||~~ ~~||||~~

mushrooms |||

extra cheese ~~||||~~   green ~~||||~~
                        peppers

squid | red ||
anchovies |||| peppers

TEACHER TIP

You might have students give 3 votes to their favorite topping, 2 votes to their second choice, and 1 vote to their third choice. Guide them to see that this method provides an even more accurate view of toppings the class favors.

Using Your Findings

Talk with students about how to use the results of the voting to figure out what toppings to choose for the party pizzas. Ask:

◆ Should all pizzas have both of the top two toppings in equal amounts?

◆ Should the pizzas have more of one topping and less of the other?

◆ Should the pizzas have only the top topping?

◆ What if someone hates *both* top toppings?

Have the class come to a consensus on the topping plan that works best.

THIRD ASSIGNMENT
Fractions

Pie in the Sky
Discuss with students how ordering a pizza involves compromise. Not only are they choosing only one or two of the many toppings available, but once chosen, the toppings go on every slice. Introduce the idea of ordering a truly ideal pie, on which each slice could have its own unique topping. For instance, why not order a 6-slice pizza with onions on 2 slices, sausage on 2 slices, red pepper on 1 slice, and green pepper on 1 slice?

Fraction Action
Discuss how to describe the special-order pizza in terms of fractions. Students might say that it is ⅓ onion, ⅓ sausage, and ⅓ pepper. Or they might say that it is ⅓ onion, ⅓ sausage, ⅙ red pepper, ⅙ green pepper. Using classifying skills, they could describe it as ⅔ vegetable and ⅓ meat.

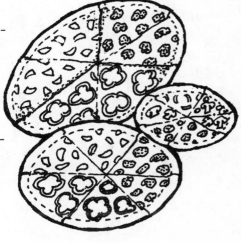

Repeat this activity with a specialized 8-slice pizza that volunteers customize. Invite students to describe the pizza in as many fractional ways as they can.

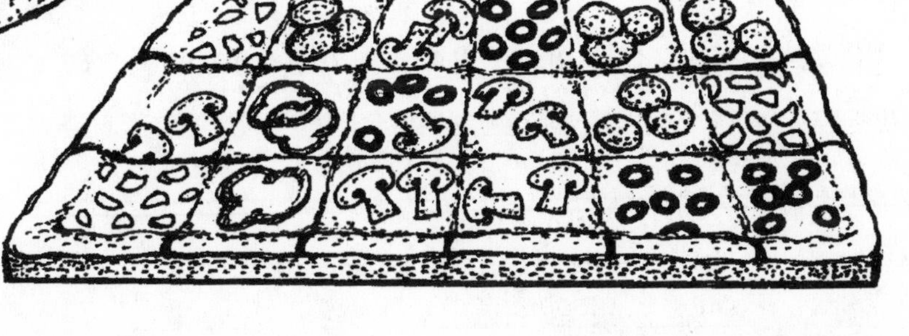

As a variation, have sudents with more experience with fractions "order" a large 18-slice party pizza that comes on a rectangular tray. They can put different toppings on different pieces of the party pizza and describe the pie in as many fractional ways as they can.

FOURTH ASSIGNMENT
Fractions
To help students explore using fractions, duplicate and pass out the worksheet on the next page.

Name_____

The Ultimate Pie

What would you put on the most original and crazy pizza you can imagine?

1. Use a ruler to turn the circle into a 6-slice or 8-slice pizza.

2. Create the ultimate pizza. Write one of your favorite toppings—peanut butter, lobster, pineapple, bubble gum, whatever!—in each slice.

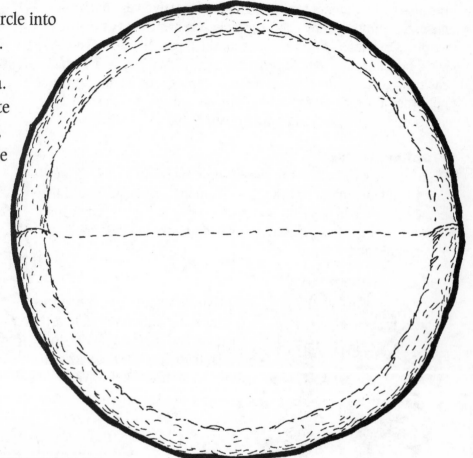

3. Give your pizza a name.

_____

4. Describe your pizza using fractions. How many different ways can you do this?

_____

_____

_____

_____

_____

_____

 **More To Try!** Write a story problem about your ultimate pizza. It might involve adding, subtracting, doubling, or any other math skill. Ask a classmate to solve your pizza problem.

FIFTH ASSIGNMENT
Calculating

How Many Pizzas?
At long last, it's time to order the food and start the feast!

To determine how much pizza the class needs, find out how many slices everyone wants. Ask students to raise one finger for each slice they'll eat at the party. Have a volunteer tally the results. Work with students to use that total to figure out how many pies to buy.

TEACHER TIP
You can order all 6-slice pies, all 8-slice pies, or a combination of both.

What Will It Cost?
Students can use local take-out menus to find the cost of each size pizza and the total cost of all pizzas needed. Have students work in small groups. Distribute different menus to each group. After students study their menu and figure the number, sizes, and total cost of the pizzas, have them compare findings to choose one restaurant to order from. Ask: Is the restaurant with the best price our best choice? Discuss factors to consider other than price.

Next, help students determine a fair pizza payment plan. To guide their thinking, ask questions like these:

◆ Will everyone pay the same price? If so, how do you determine that price?

◆ Will people pay by the number of slices they eat? If so, how do you figure out what each person pays?

◆ Have you included money to cover sales tax? A tip for the delivery person? Drinks?

Extending the Activity
Challenge students who can find the area of a circle ($A = \pi r^2$) to determine which size pizza is the best buy. Remind them to use the diameter of a pizza—10 inches, 12 inches, 14 inches, and so on—to solve this problem.

Conclude the workshop by ordering pizza and drinks, supplying napkins, paper plates, and cups, and then having the party!

TEACHER TIP
Repeat the relevant steps to determine what drinks to have at your pizza party.

What Did We Learn?
Have students look back through the activities in this workshop and jot down all the ways they used math to make decisions for the pizza party. Ask them to think about other ways they could have used math. Ask: How could this procedure be useful on other occasions? Invite students to add their responses to their math journals or portfolios.

Independent Student Activities...

...that students can do when they have time or as needed

Planning a Plot

You use measurement every day. How far is it to a friend's house? How long will it take to write a book report? How much will a new baseball cap cost? Here's another way that measurement can come in handy.

Your class is planning a community garden. You want to fence off the area where your garden will be. You have 18 feet of fence to put around your garden.

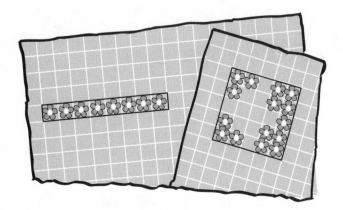

1. How many different rectangular areas can you make that have a perimeter of 18 feet? Draw each one on a piece of grid paper.

2. Find the area of each plot. Which is the largest plot? How large is it? Which plot is the smallest? What is its area?

3. How would you describe the shape of the plot with the greatest area? The least area? Which plot would be the easiest to work in? Which would be the most attractive?

Now imagine that you have a larger plot and a bigger roll of fencing—60 feet. Again, you want to fence off a rectangular area for your garden.

4. How many different rectangular plots can you make that have a 60-foot perimeter? Draw each one on grid paper.

5. What is the area of the largest plot? Which plot is the smallest? How small is its area?

6. How would you describe the shape of the plot with the greatest area? The least area?

7. What generalization can you make about the relationship between the shape of a rectangle and its area?

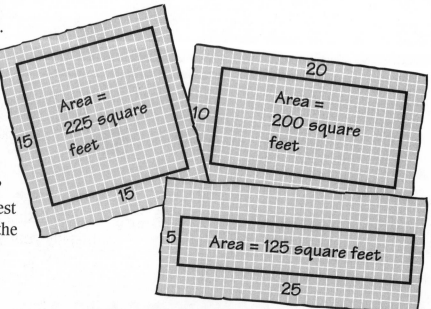

Tessellations

A tessellation is an arrangement of one or more repeating shapes. These shapes completely fill a plane without overlapping or leaving any open space. Designers use tessellations in repeating patterns on floor tiles or fabric. Tessellations often appear in the mosaic tile patterns of Islamic art.

Shapes That Tessellate

Work with a partner. Create designs by drawing repeated polygons on grid paper to see which shapes can completely cover a section of the page. (Or use pattern blocks to try to cover your desk.) Which polygons tessellate?

You can create your own tessellation designs. Here's how:

1. On oak tag, draw a shape that tessellates, such as any quadrilateral.

2. Cut out your shape and place it on a sheet of paper. Trace around it.

3. Slide the shape to a new position, where one side touches a side you traced.

4. Trace the shape in its new position. Continue in this way—slide and trace, slide and trace—to make a design. Remember not to leave any empty space between tracings.

5. Color your tessellations to highlight the design.

Adaptation Tessellations

You can adapt shapes that tessellate to make new tessellating shapes. Here's what to do:

1. Cut from oak tag a shape that tessellates. Cut out part of your shape.

2. Reattach the part with tape somewhere else on the remaining shape to create a new shape. Do not overlap the edges.

3. As before, trace and slide the new shape to create a design.

Here's an example:

4. Now try it yourself. Create a shape. Cut and rearrange it, trace and slide it to form a tessellating design, and color the design. Display your adaptation tessellation in the classroom, along with the shape you used to create it.

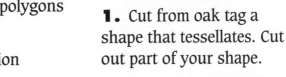
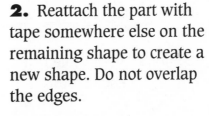

What a Day!

Every pet has its day. But what exactly does a pet do all day? Make a table to show how your pet (or a pet you imagine) spends its day. Then display the data from the table in a circle graph.

1. Begin by listing all the animal's daily activities in a table like this. Keep in mind that, just like your day, a pet's day has 24 hours.

SPOT'S TYPICAL DAY	
ACTIVITY	TIME SPENT
sleeping	12 hours
eating	1 hour
chewing on clothes	4 hours
running and fetching	3 hours
watching you do homework	4 hours

2. Use the information in your table to make a circle graph. Label each sector of your graph with the pet's activity and the number of hours spent at it.

Hint Think of each amount of time as a fraction of 24 hours. For example, if your pet exercises for 6 hours, the sector (pie slice) of your graph that shows this will be $\frac{6}{24}$ or $\frac{1}{4}$ of the circle.

3. Write a fraction to represent the part of the day your pet spends doing each of its activities. Remember to express all fractions in their simplest form.

4. Use the information you gathered to answer these questions.

◆ How would you summarize the way your pet spends its day?

◆ What fraction of the day does your pet spend doing what it does the most? The least?

◆ What, if anything, does your pet do for $\frac{1}{6}$ of a day? For $\frac{1}{4}$ of a day?

◆ How does your pet's day compare with yours?

5. Examine a classmate's graph for the same kind of pet you have. Compare and contrast how the two pets spend their days. Use fractions in your answer.

Shopping at a Fraction of the Cost

Knowing how to compute and estimate with fractions can turn ordinary shoppers into smart shoppers. Shoppers who know their fractions can save a bundle at one of Big Sal's big sales!

Suppose a popular video that usually costs $23.95 is on sale at ¼ off. Here's how to estimate the sale price and the savings:

◆ Round $23.95 to $24, a number that's easily divisible by 4.

◆ Multiply to find ¼ of $24: ¼ × $24 = $6; You save $6.(Or you can divide, since ¼ means 4 equal parts: $24 ÷ 4 = $6.)

◆ Subtract: $24 – $6 = $18; The sale price will be about $18.

Use the information in Big Sal's ad to answer the questions.

BIG SAL'S BIGGEST SALE

ITEM	LIST PRICE	DISCOUNT
Hockey Jersey	$20.00	⅕ Off
Exercise Mat	$29.79	⅓ Off
Big Sal T-shirt	$15.95	¼ Off
Backpack	$21.29	⅓ Off

1. How much do you save by buying the hockey jersey on sale?

2. About how much will the exercise mat cost on sale? How about the T-shirt?

3. At another store, the same backpack usually sells for $24, but it's on sale there for ½ off. Which is the better buy—the backpack on sale at Sal's or at the other store?

Dozens of Discounts

Work with a partner. Make up a sale flyer for a store of your choice. List 4–6 items, showing what each usually costs. Give a discount for each item during the sale. Write the discount as a fraction. Write some shopping questions to go with your data. Then swap flyers and questions with another pair of smart shoppers.

Solid Thinking

Imagine that you are designing a platform for a tent. Naturally, you need to make sure the tent won't collapse! So you must make the columns that support it as strong as possible.

Guess and Test

What shape should the columns for your platform be? Should they be cylinders, squares, or triangles?

First, write down your guess. Then do an experiment to find out. You'll need three index cards, tape, and some paperback books.

1. Make a cylinder by taping together the two short sides of an index card. Let the sides overlap a little.

2. Make a square column from another index card. First make a very narrow fold parallel to one of the short sides to form a tab. Fold the rest of the card in half so that the other short side touches the first fold. Fold each half in half. Then fold the card into a square column. Tape it in place, using the narrow fold as a tab.

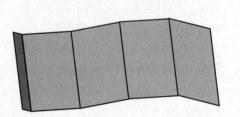 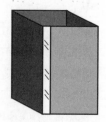

3. Make a triangular column from the third index card. First make a very narrow fold parallel to a short side to form a tab. Then fold the rest of the card into thirds. Fold the card into a triangular column and tape down the tab.

4. Now test the strength of each column. Carefully stack paperbacks, one at a time, on top of one of the columns. Add books until the column collapses. Record how many books the column supported before it collapsed. (If the column falls over because of bad balance, try again.) Repeat for the other two columns: stack and record.

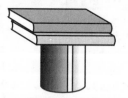

Which shape made the strongest column?

Check It Out

Look at the columns used on buildings in your community. Note their shapes. Which shape is most commonly used? Why do you think that is?

Stem-and-Leaf Plot

A stem-and-leaf plot is one way to organize and display data.

PLAYER	AGE
Smith	38
Jones	29
Torrez	32
Reilly	18
Cohen	31
Tortelli	41
Fisher	22
Tsang	32
Gonzales	31
Steen	24

The table and the stem-and-leaf plot below show the same information about the ages of players on a baseball team:

STEM	LEAF
1	8
2	2 4 9
3	1 1 2 2 8
4	1

Compare the stem-and-leaf plot with the table. Notice that the data on the plot are ordered, as in a frequency table. The front-end digits—the tens digits 1, 2, 3, 4—make up the stems. The leaves are ones digits.

So the player named Reilly is shown as

STEM	LEAF
1	8

Fisher's age is shown as

STEM	LEAF
1	8
2	2

Steen's age is shown as

STEM	LEAF
1	8
2	2 4

Jones is 29. How is his age given?

STEM	LEAF
1	8
2	2 4 9

Data in a stem-and-leaf plot are shown visually, as in a bar graph. For example, it is easy to see that more players are in their thirties than in any other age group, because the "line" formed by the leaves is longest in the row with the stem of 3. What else does the stem-and-leaf plot show you about the players' ages?

Making a Stem-and-Leaf Plot

1. Use a tape measure to measure the "wingspan," in inches, of at least 10 of your classmates. List each person's name and his or her "wingspan" measurements on a table.

2. Label two other columns **Stem** and **Leaf.**

3. Look at the front-end digits of the measurements in your table. Use these as the stems. List them in the Stem column.

4. Copy each ones digit leaf onto the table beside its corresponding stem. Arrange the leaves in each row from least to greatest.

Use your completed stem-and-leaf plot to answer these questions.

1. What does the plot show about the lengths of your classmates' wingspans?

2. Which wingspan appears most often? (This statistic is called the *mode*.) How does the stem-and-leaf plot make the mode easy to see?

3. What are the longest and shortest wingspans? How does the plot show this?

Pack Your Calculator

If you're traveling abroad, it's a good idea to pack your calculator. Different countries have different money systems. When you travel to another country, you exchange American dollars for the money of that country. Usually, the value of a dollar is different from the unit of another country's money. You have to find out what the exchange rate is. This is easy to do on a calculator.

Read the Rates

The table shows the value of American dollars in terms of some foreign money. It shows how many of each unit of foreign money you get in exchange for $1. For example, $1 = 32.25 Indian rupees. So an American visitor to India will get 32.25 rupees for each dollar. To find the total number of rupees for a trip to India, a traveler multiplies the number of dollars he or she has by 32.25.

An Indian who plans a visit to the U.S. must exchange rupees for dollars. He divides the number of rupees he has to spend by 32.25 to find out how many dollars are equivalent to it.

Study the foreign exchange rates shown on the table. Then use a calculator to answer the questions that follow.

Country	1 Dollar =
Brazil	0.908 real
China	8.28 yuan
France	4.91 francs
India	32.25 rupees
Saudi Arabia	3.75 riyals

1. Which is worth more dollars, 100 rupees or 100 reals? How can you tell?

2. Suppose an American girl has $100 to exchange. Would she get more francs or riyals for her money? How do you know?

3. Mr. Sandoz is going to China on business. He brings $1,000 to spend. How many yuan will he get in exchange for his dollars?

4. From China, Mr. Sandoz goes to Paris, France. He has $500 when he arrives there. How many francs can he get in exchange?

5. Marie flies to Chicago from France to visit relatives. She has 500 francs to spend. How many dollars will she get in exchange?

What's It Worth?

Many newspapers provide information on money exchange rates. You can also find this information at some banks. Find out what the exchange rate is for some other countries. Notice that this rate changes daily. Keep a record for a week. On what day did American travelers get the best rate?

Number Words Around the World

Many people travel for business or pleasure. When you visit another country, it is helpful to know some key words in the language of that country. Number names are especially useful. Can you think of reasons why?

Study the chart on this page. It shows you how to count in four languages.

Name the Number

Work with a partner to use the number names on the chart. Make up four questions to ask your partner. The answer for each question must be a number from 1 to 10. For example: How old are you in Swahili?

Write a Word Problem

Write simple word problems using the number names from one of the languages on the chart. For example: Greta had acht carrots and she gave zwei to her friend Wilt. How many carrots did Greta have left? *(sechs)*

Number Name Challenge

Can you count to ten in French? Italian? Japanese? Find out the names for the numbers 1 to 10 in at least two other languages. Make a new chart to show these words. Or work with classmates to make a large classroom chart with all the languages you have learned.

Number	Spanish	Swahili	German	Chinese
1	uno (oo-noh)	moja (mo-jah)	eins (eye'ns)	一 (yee)
2	dos (dohs)	mbili (mm-bee-lee)	zwei (ts'vy)	二 (uhr)
3	tres (trehs)	tatu (tah-too)	drei (dry)	三 (sahn)
4	cuatro (kuah-tro)	nne (nn-nay)	vier (feer)	四 (suh)
5	cinco (sin-co)	tano (tah-no)	funf (finf)	五 (woo)
6	seis (seh-iss)	sita (see-tah)	sechs (seks)	六 (lyo)
7	siete (syeh-teh)	saba (sah-ba)	sieben (see-ben)	七 (chee)
8	ocho (o-cho)	nan (nah-nay)	acht (aht)	八 (bah)
9	nueve (n'weh-veh)	tisa (tee-sah)	neun (noin)	九 (jo)
10	diez (d'yess)	kumi (koo-mee)	zehn (tsehn)	十 (shur)

Grid Logic

When you are solving problems, sometimes it's helpful to use visual aids. Sarah, Pedro, Art, and Naomi play on different baseball teams. Here are some clues to help match each person with his or her team.

Try the following logic problem about baseball players and caps. Make a grid. Use checks to record information you know is correct about players and their caps. **HINT:** It can help to eliminate choices you learn are not correct. Use a different mark to show mismatches.

◆ Naomi is *not* on the Rockets.

◆ Sarah's team is named after a kind of animal.

◆ Pedro's name starts with the same letter as his team.

You can use a grid to solve logic problems like this. Make checks to indicate facts you know. In the grid for the baseball problem, checks match each player with his or her team. Show mismatches with X's. From these facts, you could conclude Naomi must be on the Hurricanes and Art is on the Rockets.

Yasmina, Gary, Sam, and Rita are on different baseball teams. Each team has a different color cap—red, silver, green, and yellow. Use the clues to match the players with their cap colors.

◆ Yasmina and the boy with the silver cap live in the same apartment building.

◆ Sam rides a bus from his house to Yasmina's neighborhood to practice.

◆ Yasmina's cap is *not* yellow.

◆ Rita has a green and purple team jacket to match her cap.

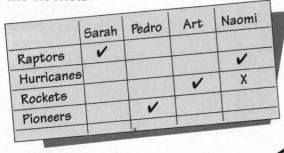

	Sarah	Pedro	Art	Naomi
Raptors	✔			✔
Hurricanes			✔	X
Rockets				
Pioneers		✔		

Writing Math Stories

Some say a picture is worth a thousand words.
Well, it's also worth several math problems.

Get the Picture
Study the picture. What fraction of the skaters are fixing their skates? What fraction are wearing headphones?

Now it's your turn to use the picture to make up math problems.

1. Write a problem that involves adding fractions.

2. Write a problem that involves multiplication.

3. Write a problem that involves both multiplication and subtraction.

Be sure to ask a classmate to solve your math stories.

Skills-Practice Reproducibles...

...to use a just the right moment and in your math center

Name_____

Straight Lines That Curve

When do straight lines curve?
When you draw them in a parabolic pattern.

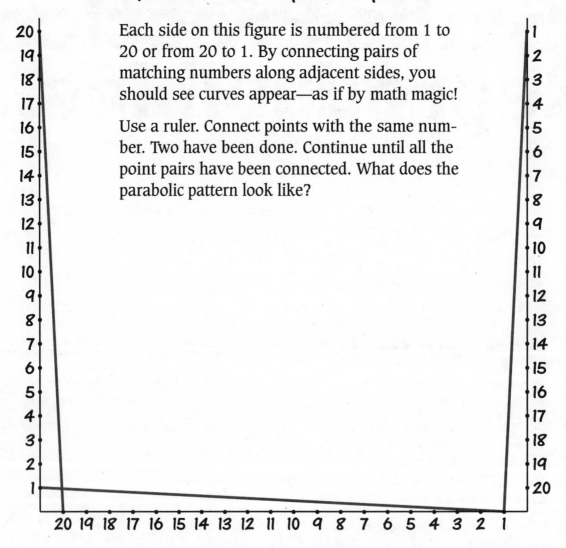

Each side on this figure is numbered from 1 to 20 or from 20 to 1. By connecting pairs of matching numbers along adjacent sides, you should see curves appear—as if by math magic!

Use a ruler. Connect points with the same number. Two have been done. Continue until all the point pairs have been connected. What does the parabolic pattern look like?

You can make a beautiful piece of math art by doing parabolic patterns inside various shapes, or by using colored pencils. A beautiful alternative is to use colored yarn to stitch points together.

More To Try! Try some of your own parabolic patterns on grid paper. Try straight-line curves inside triangles or other shapes. Use a ruler to draw the outline. Number the sides in alternating order (low to high, then high to low). Connect the points as before.

Name_____

Time for Cartoons

You are planning an all-morning cartoon festival. Your goal is to schedule cartoons to be shown, in any order, to fill the time from 9:00 A.M. to noon.

The schedule must follow these rules:

◆ You must show at least 5 cartoons.

◆ Leave 3–5 minutes between each cartoon.

◆ The cartoons must begin on times that are multiples of 5 (like 10:40, not 10:42).

The cartoons listed at right have been chosen by the selection committee. Use the information given to make your schedule.

Cartoon Title	Running Time
Marvin the Moose	22 minutes
Ghouls on the Go, Part 3	34 minutes
Large Invaders From Iceland	26 minutes
Bears on Bikes	14 minutes
Space Station Rodeo	21 minutes
Underwater Rabbit Returns	38 minutes
Lost Cities on Jupiter	17 minutes
Jungle Mouse and the Big Cheese	20 minutes

Make your schedule using a table like this.
An example is given to show you how to complete the chart:

CARTOON	START TIME	END TIME
Space Station Rodeo	9:00 A.M.	9:21 A.M.
Bears on Bikes	9:25 A.M.	

 **More To Try!** Describe how you would adjust your schedule if you needed a 10-minute break between cartoons.

Name_____

Share and Repair

You can use your math and thinking skills to solve many different kinds of problems. Here are two to try.

1. Eight people are planning to share a storage room in a warehouse. Each person needs an equal size area. Can you divide the storage room so that everyone has the same size space? Draw lines to show how you would do this.

2. A man drops a mirror and the glass breaks into four pieces. Can you help the man fit the pieces back into the frame? Trace the pieces and draw them within the frame.

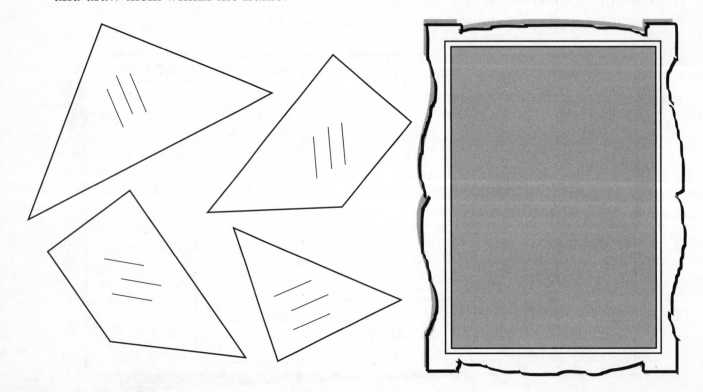

More To Try!

After six months, four people decide to take their things out of the storage room. Can you divide the room equally for the remaining four people?

Name_____

Math on a Map

**Maps can tell you how far apart places are.
You can use this information to figure out travel plans.**

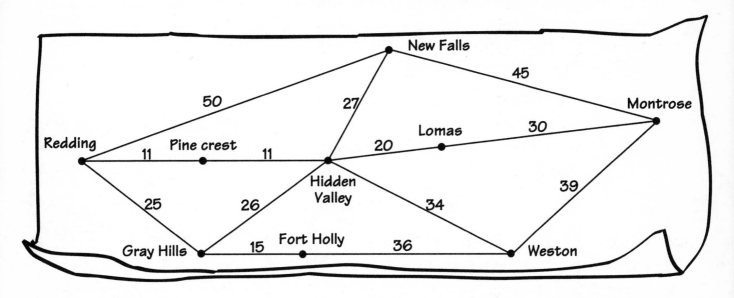

Study this map. The numbers tell how many miles it is from one place to another.
Use this information to answer the questions.

1. What is the quickest way to get from Weston to New Falls? _____

2. Is Hidden Valley closer to Redding or to Lomas? _____

How much closer is it to Pinecrest than to Redding?_____

3. Which route is shorter: from Lomas to New Falls or from Lomas to Redding?

_____

4. If you were traveling from Montrose to Fort Holly, which would be the shortest route?

_____

5. What is the shortest route from Pinecrest to Gray Hills? _____

How much shorter is this route? _____

**How many miles is it to go from Montrose to Redding with a
stop in Fort Holly?**

Name_____

A-Maze-ing True Facts

Real life math uses many relationships that are equivalent—always equal. For instance, 1 week always has 7 days. Do you know enough math equivalents to find your way through this maze? Try it and see!

Object: Make a connected path on math equivalents from START to FINISH.
A correct space has a true math equivalent in it.

Moves: Begin at START. Put a counter on any space that touches START horizontally or vertically. You may NOT move diagonally (corner to corner). Keep making your way toward FINISH along true math equivalents. Always follow the rules of touching spaces.

WATCH OUT! Don't be tricked by false equivalents.

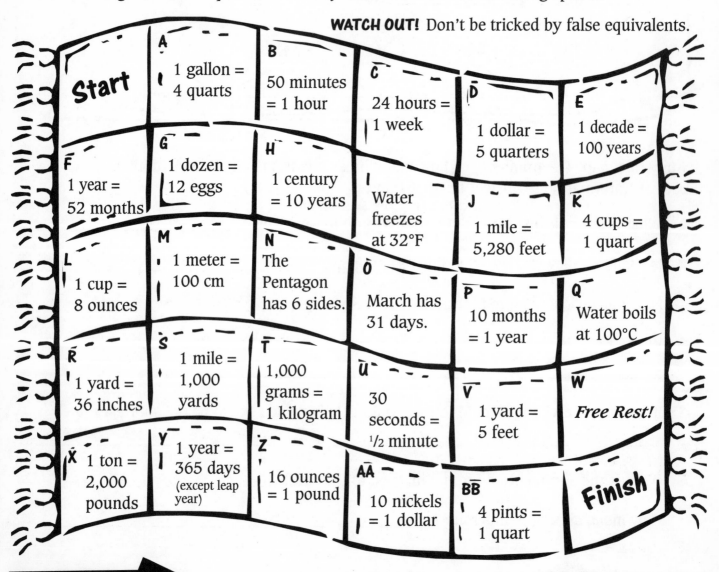

	A	B	C	D	E
Start	1 gallon = 4 quarts	50 minutes = 1 hour	24 hours = 1 week	1 dollar = 5 quarters	1 decade = 100 years
F 1 year = 52 months	**G** 1 dozen = 12 eggs	**H** 1 century = 10 years	**I** Water freezes at 32°F	**J** 1 mile = 5,280 feet	**K** 4 cups = 1 quart
L 1 cup = 8 ounces	**M** 1 meter = 100 cm	**N** The Pentagon has 6 sides.	**O** March has 31 days.	**P** 10 months = 1 year	**Q** Water boils at 100°C
R 1 yard = 36 inches	**S** 1 mile = 1,000 yards	**T** 1,000 grams = 1 kilogram	**U** 30 seconds = 1/2 minute	**V** 1 yard = 5 feet	**W** *Free Rest!*
X 1 ton = 2,000 pounds	**Y** 1 year = 365 days (except leap year)	**Z** 16 ounces = 1 pound	**AA** 10 nickels = 1 dollar	**BB** 4 pints = 1 quart	**Finish**

More To Try! Make up your own math game for two people to play that uses true and false math equivalents. Play it with a friend.

Name_____

DOT'S THE ANSWER

In this game you practice some math skills and complete a picture at the same time.

Complete the math problems on the page. Check your answers. Then put your pencil on the arrow on the picture. Trace a line from dot to dot using the answers in row 1 (from A to G) as a guide. When you have connected the answers from row 1, do the same with row 2.

	A	B	C	D	E	F	G
1	432 × 3	590 × 1	221 × 2	678 × 5	943 × 8	209 × 3	145 × 4
2	843 × 7	255 × 6	901 × 3	734 × 2	103 × 8	654 × 9	277 × 2

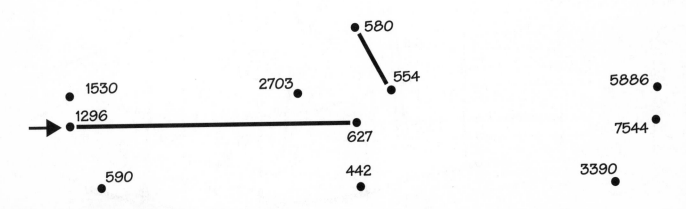

• 5901

1468 •

• 824

• 580

• 554

1530 •

2703 •

1296 →———————• 627

5886 •

7544 •

• 590

442 •

3390 •

More To Try! Write a caption for your picture.

Name_____

Math Mistakes

Often math mistakes can help you learn. But some math mistakes can cause real problems. Look at the picture. The people in this town have made a lot of math mistakes. Can you find them?

Find 10 mistakes. Correct them with a red pencil.

BAKERY

Hours
8 PM
to
8 AM

TOYS

We're open 8 days a week

PARKING $25. HR

Clothing

25% OFF!

SWEATERS REDUCED FROM $30. TO $15

THE FOUR LEAF Clover

ICE CREAM

A DOZEN GREAT FLAVORS

VANILLA   PEACH

LEMON   MINT

MOCHA   BERRY   ALMOND

MAPLE   COCONUT

CHOCOLATE   CHERRY

THE Music Room

SALE
All books
$ .495

BOOKS

NOW PLAYING:
The
BUZZ
BUSTERS
JUNE 15-
JUNE 31

TONIGHT!
The
Barbershop
Quartet

SPEED
LIMIT
30
MINUTES
PER
HOUR

More To Try! On another sheet of paper draw your own picture with math mistakes. Ask a classmate to find the errors.

Name_____

Math-squerade

You're having a costume party. You've told everyone to wear a funny
face disguise. Around your house, you find these possibilities for
your own funny face.

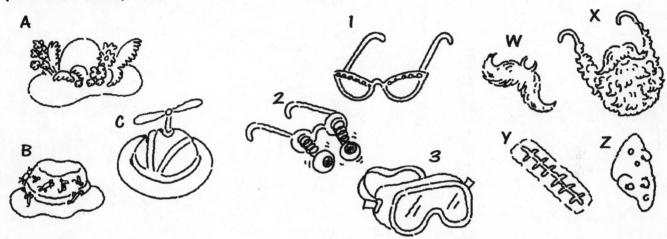

Answer these questions about funny face combinations for the party.

1. A disguise must have 3 parts: 1 hat, 1 type of glasses, and 1 type of face changer.
Which parts would you choose to make your disguise? _____

_____

2. Suppose you're sure that you'll wear the eyeball glasses.
How many different disguises can you make? _____

3. List every possible disguise that uses eyeball glasses. _____

_____

_____

4. How many possible different disguises are there in all? Make a list on another page.
Remember, every disguise needs a hat, some kind of glasses, and some kind of face
changer.

5. What relationship do you notice between the number of items in each group and the
total number of funny face disguises you can make? _____

_____

 Make up your own groups of things for a silly disguise. Ask a
friend to find out how many different disguises are possible.

71

Teacher's Notebook...

...including tips on assessment, technology, and a bibliography of fiction and nonfiction for students and resources for teachers

Portfolio and Assessment Ideas

From journal writing to storytelling, your students may be involved in a wide range of activities as they explore and learn math skills and concepts. Here are suggestions for assessing and evaluating their understanding.

File It Fast

Anecdotal observations offer ongoing insight into a child's mathematical reasoning and understanding that answers alone cannot provide. Make sure those valuable comments get filed with these record-keeping tips.

Self-Stick Savers Do you use an index card filing system to record observations? Instead of pulling each child's card to make notes throughout a lesson (and then ending up with cards all over your desk because there's no time to refile), keep self-stick notes on hand. When you see or hear something you'll want to look back on, just date a note and write the child's name and a remark. At the end of the day or week, stick the notes on students' cards. Hint: Use big index cards or, better yet, file folders.

Kid Calendars Create a simple calendar template, showing only weekdays. Before making copies for each child, write in the month and dates (or ask students to fill in their own). Punch holes in each calendar and place in a three-ring binder. As you observe students at work, you can easily record comments for each child on the appropriate page and space. Make a fresh set of calendars at the start of each month. These day-to-day records of observations will give you an organized, easy-to-consult picture of progress over time.

The Stamp of Organization

Portfolios are a powerful way to showcase student progress. But keeping everything organized so that you can see the progress is another matter. A child's collection of math work might include:

- ◆ projects
- ◆ models
- ◆ graphs
- ◆ drawings
- ◆ word problems written by students
- ◆ journal entries
- ◆ tape recordings of explanations
- ◆ descriptions of problem-solving strategies and other written responses

A date stamp can help students manage these growing collections of work. Just have students stamp any completed work that they want to keep in their portfolio collections.

Before, During, After

Involve students in their own assessment by creating before, during, and after forms. Before a new math activity, ask students to write down something they know or wonder about the concept or problem. During the activity, remind students to complete the middle column, writing about their thinking or about problems they're having. After completing the activity or lesson, ask students to write something they've learned.

DATE	ACTIVITY	BEFORE	DURING	AFTER

Calculator Connections

TEACHER TIP
You'll find another calculator activity on page 58.

The calculator is an inexpensive, handy, and versatile technological tool. Here are some ways to use calculators to bring real-life math to the classroom.

Fractions to Decimals

Most calculators cannot operate with fractions, so real-world situations involving fraction computations require fractions to be changed to decimals. Calculators also help students learn the relationships between fractions and their decimal equivalents. Demonstrate how to use a calculator to divide the numerator by the denominator. Enter $3 \div 4 =$ to show that $\frac{3}{4} = 0.75$.

Balance the Checkbook

Many households pay bills by check, so it's important to keep track of how much money is in the account at any given time. Prepare a mock check register for students to use. Include payments, deposits, and an opening balance. Have them use calculators to find the running balance.

Patterns

Number patterns are fun to explore with calculators. Present these patterns to students. Have them use calculators to find the first three products and describe the pattern they see. Have them predict the fourth product, then verify it with the calculator.

$5 \times 5 =$	$37 \times 33 =$	$101 \times 101 =$
$5 \times 55 =$	$37 \times 333 =$	$101 \times 202 =$
$5 \times 555 =$	$37 \times 3333 =$	$101 \times 303 =$
$5 \times 55555 =$	$37 \times 33333 =$	$101 \times 707 =$

Binary Investigation

Computers operate on binary code, using zeros and ones to represent every character or number. Have students think like computers to form numbers by entering *only* zeros and ones in their calculators, as well as operations signs (+, −, ×, ÷) and =. To make 342, **enter 100 + 100 + 100 + 10 + 10 + 10 + 10 + 1 + 1.**

Constant Counting

Show students how to use the constant function to skip count. In most calculators, the constant function is in the equals key. To skip count by 4, **enter 4 + 4 = = =...** The calculator will show 8, 12, 16... as many times as the equals key is pressed. Students can skip count forward or backward.

Weird Statistics

Students can use calculators to create their own fantastic stats. For instance, to estimate how many times you've breathed since birth:

1. Figure out a breath rate for 1 minute.

2. Find how many breaths in 1 hour (multiply by 60).

3. Find how many breaths in 1 day (multiply by 24).

4. Find how many breaths in 1 year (multiply by 365).

5. Find how many breaths in the *n* years they've been alive (multiply by *n*). For an eleven-year-old: $16 \times 60 \times 24 \times 365 \times 11 = 92,505,600$ breaths!

Bibliography

Nonfiction

Anno's Mysterious Multiplying Jar by Masaichiro and Mitsumasa Anno (Putnam)

Circles by Catherine Sheldrick Ross (Scholastic)

Clocks: Chronicling Time by A.J. Brackin (Lucent Books)

Computers by Karen Jacobsen (Childrens Press)

Computers in Action by Nigel Hawkes (Watts)

Computers in Your Life by Melvin Berger (Harper)

Count Your Way Through Africa; Count Your Way Through the Arab World; Count Your Way Through China; and *Count Your Way Through Israel* by Jim Haskins (Carolrhoda Books)

The Division Wipe-Off Book and *The Multiplication Wipe-Off Book* by Alan Hartley (Scholastic)

Everything You Need to Know About Math Homework by Ann Zeeman and Kate Kelly (Scholastic)

Everything You Need to Survive: Money Problems by Jane Stine (Random House)

Exploring Triangles: Paper Folding Geometry by Jo Phillips (Harper)

Get the Message by Gloria Skurzynski (Bradbury Press)

How Much Is a Million? and *If You Made a Million* by David M. Schwartz (Scholastic)

I Hate Mathematics Book by Marilyn Burns (Scholastic)

The Kid's Complete Guide to Money by Kathy S. Kyte (Knopf)

The Kid's Money Book by Patricia Byers (Liberty)

Margaret's Moves by Berniece Rabe (Scholastic)

Math Mini-Mysteries by Sandra Markle (Atheneum)

263 Brain Busters: Just How Smart Are You Anyway? by Louis Phillips (Viking)

Young Fu of the Upper Yangtze by Elizabeth Lewis (Henry Holt)

Fiction

Adam of the Road by Jane Gray (Scholastic)

Anastasia on Her Own by Lois Lowry (Dell)

The Boggart by Susan Cooper (McElderry Books)

Chip Rogers, Computer Whiz by Seymour Simon (Morrow)

Danny Dunn and the Homework Machine by Jay Williams (McGraw-Hill)

The Door in the Wall by Marguerite de Angeli (Dell)

Freckle Juice by Judy Blume (Macmillan)

How Big Is a Foot? by Rolf Myller (Dell)

The Hundred Penny Box by Sharon Bell Mathis (Scholastic)

The King's Chessboard by David Birch (Dial)

Mail Order Wings by Beatrice Gormley (Dutton)

Peter Graves by William Pene Du Bois (Puffin)

Rosy Cole's Great American Guilt Club by Sheila Greenwald (Little, Brown)

Periodicals

Zillions magazine (Consumers Union)

Teacher's Resources

Estimation Investigations and Great Graphing by Marcia Miller and Martin Lee (Scholastic)

Garbage Pizza, Patchwork Quilts, and Math Magic by Susan Ohanian (Freeman)

Literature-Based Math Activities by Alison Abroms (Scholastic)

The Man Who Counted by Malba Tahan (Norton)

Math All Year by Linda Ward Beech (Sniffen Court Books)

Multicultural Math by Claudia Zaslavsky (Scholastic)

Real World Math by Karen Brown (Edupress)

Thematic Book Reports for Math by Linda Milliken (Edupress)

30 Wild and Wonderful Math Stories by Dan Greenberg (Scholastic)

Using Calculators Is Easy! by Char Forsten (Scholastic)

Answers to Student Worksheets

Page 11
Answers will vary.

Page 17
Answers will vary.

Page 42
Check to be sure students have followed the directions to make a seating plan.

Page 45
Students' glyphs will vary.

Page 48
Students' pizzas will vary.

Page 64
Check to be sure students connect the point pairs to make a parabolic pattern.

Page 65
Schedules will vary.

Page 66

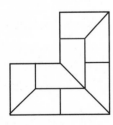 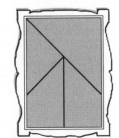

Page 67
1. through Hidden Valley
2. Lomas; 11 miles
3. Lomas to Redding
4. through Weston
5. through Redding; 1 mile

Page 68
A, G, M, L, R, X, Y, Z, T, U, O, I, J, K, Q, W

Page 69
Check to see that students draw a sailboat.

Page 70
Possible corrected mistakes:
   8 AM to 8PM
   Parking 25¢ per Hour
   Open 7 Days a Week
   Sweaters 25% off reduced from $30 to $22.50
   Add a leaf to make four-leaf clover
   Add an ice cream flavor to make a dozen
   Remove one face from Barbershop Quartet
      picture
   Speed Limit of 30 miles per hour
   June 15–June 30
   All books $4.95.

Page 71
1. Disguises will vary.
2. 12
3. 2+A+W, 2+A+X, 2+A+Y, 2+A+Z; 2+B+W, 2+B+X, 2+B+Y, 2+B+Z; 2+C+W, 2+C+X, 2+C+Y, 2+C+Z
4. 36.
5. Students should note that the disguise combinations are multiples of the disguise elements.

Notes

Notes

Notes